Sustainable Development of Ecosystem, Wildlife and Heritage Conservation for Human Welfare

The Editor

Dr Ashwani Kumar Dubey is President of Environment and Social Welfare Society, Khajuraho and Editor-in-Chief of "International Journal of Global Science Research, India". He is currently working as Officer, Information Technology, Maharaja Chhatrasal Bundelkhand University, Chhatarpur, Madhya Pradesh, India.

He did M. Sc. in Zoology with specialization of Ichthyology in 1991 from Awadhesh Pratap Singh University, Rewa, Madhya Pradesh. He obtained his Ph. D. Degree in 1995 entitled "Responses of Antioxidants, Lipid-protein Interactions and Lipid Peroxidation in *Heteropneustes fossilis* to oxidative damage exposure" from School of Studies in Zoology, Vikram University, Ujjain, Madhya Pradesh. His research field was Biochemistry, Free Radical Biology, Toxicology and Stress Monitoring. He has published several research papers in Foreign Journals including Pergamon press UK and USA to his credit. He published many popular science articles, Delivered Lectures and broadcasted Science Talk by All India Radio for the Student and Social Welfare.

He joined as a Scientist, Research and Development in the Rank Industries Ltd., Nellore, in Andhra Pradesh State in 1995. During this service period he served for Pathological and Water Quality Assurance in Aquaculture Division. Then in 1998 he was appointed as an Assistant Professor of Zoology at RBS College, Rajnagar. He managed the College through a major transformation of its research and teaching. He played a major role in the design and construction of a new Fisheries Demonstration Centre at Godavari Estate, Nahdora-Khajuraho, India. In 2004 he became Guest Lecturer of Zoology, in Higher Education Department, Government of Madhya Pradesh. For most of his career his research interests have focused on the Biodiversity, Environmental Impact Assessment and Bio-Resources Conservation.

Under the able guidance of Executive Director Dr. Ashwani, the ESW Society has curved the niche through various activities, programs and awareness camp to promote the case of Environment and social welfare. Apart from this, the Society established Godavari Academy of Science and Technology in 2010 to promote scientific awareness for environmental conservation awareness through seminar, workshop, conference and symposium. To support the case of environment awareness through scientific thinking and reasoning, the society has started International Journal of Global Science Research (IJCSR) providing a huge platform for academic interaction through research papers and analytical interpretation.

Dr. Ashwani awarded many prestigious awards by National and International institution. He is a regular member and fellow of important scientific societies and also in editorial board member Research Journals in India, America, United Kingdom, Egypt, France, Romania, Syria, Nepal and Japan. And he is an Advisor of Research Board of America, USA.

Currently Awarded by Editorial Excellence Award-2017 by I2OR India.

He has devoted his life in Academic and Scientific research because of not having fulltime employment. He is currently working on Impact of environmental degradation, global health, natural resource, wildlife conservation and stress monitoring.

His personal interests include Reading, Writing, Traveling and Photography.

Address: Dr. Ashwani Kumar Dubey, Godavaripuram, Ward No. 17, Chhatarpur 471001 India

Email: ashwanikhajuraho@gmail.com **Website:** http://godavariacademy.com

Mobile: 09425143654

Sustainable Development of Ecosystem, Wildlife and Heritage Conservation for Human Welfare

– Editor –

Dr Ashwani Kumar Dubey

2018

Daya Publishing House®
A Division of

Astral International Pvt. Ltd.
New Delhi – 110 002

ISBN 9789388173711 (Int. Edn.)

Publisher's Note:

Published by	: **Daya Publishing House®**
	A Division of
	Astral International Pvt. Ltd.
	– ISO 9001:2015 Certified Company –
	4736/23, Ansari Road, Darya Ganj
	New Delhi-110 002
	Ph. 011-43549197, 23278134
	E-mail: info@astralint.com
	Website: www.astralint.com
Digitally Printed at	: **Replika Press Pvt. Ltd.**

Acknowledgement

This is an honor for Environment and Social Welfare Society, Khajuraho, organize its ESW Vth Annual National Research Conference on **"Sustainable development of Ecosystem, Wildlife and Heritage conservation for Human welfare 30 and 31 January, 2018** at UNESCO world heritage site Khajuraho of India, Assisted by Godavari Academy of Science and Technology, Chhatarpur, MP.

I am Thankful to Secretary, Bundelkhand Extended Region Chapter, Chitrakoot, The National Academy of Sciences India, Allahabad, UP, and to Vice Chancellor, Maharaja Chhatrasal Bundelkhand University, Chhatarpur MP for its in association with ESW Society for organizing this Conference.

It is my privilege and pleasure to express my profound gratitude to our **VIP Guest** Honourable **Prof. P. K. Verma,** Vice Chancellor, Barkatullah University, Bhopal, Madhya Pradesh, **Prof. K. K. Sharma**, Former Vice Chancellor, MDS University Ajmer, Rajasthan and **Prof. Prakash S. Bisen**, Former Vice Chancellor, Jiwaji University, Gwalior, who have given very kindly, consented for Inaugural Programme of ESW Conference.

Honourable **Mrs. Lalita Yadav**, State Minister, Government of Madhya Pradesh, Honourable **Kun. Vikram Singh**, MLA, Rajnagar Vidhan Sabha, **Dr. Niraj Kumar**, Executive Secretary, National Academy of Sciences India, Allahabad, **Dr. K. S. Tiwari**, Regional Director, Indira Gandhi National Open University, Bhopal, **Dr. Shivesh Pratap Singh,** Secretary, Bundelkhand Extended Region Chapter Chitrakoot, NASI, Allahabad, Uttar Pradesh who have given very kindly, consented for Award ceremony of ESW Conference.

I am heartily thankful to honorable Invitee Guest Who have very kindly consented and given us an opportunity to share valuable thought which will provide milestone on the way of leading Scientists in the Conference.

I am especially thankful to all delegates who actively participated in this Conference. I am thankful to Electronic and Print Media. I am profoundly thankful to my Board of Director and All members of ESW Society for their invaluable cooperation, and those entire person who are directly or indirectly concerned with this conference.

Dr. Ashwani Kumar Dubey

Dr Ashwani Kumar Dubey
Executive Director
Environment and Social Welfare Society, Khajuraho
Advisor
Research Board of America

Website: www.godavariacademy.com
Email: ashwanikhajuraho@gmail.com
Tweeter: tweeter/ashwanikumardub
Mobile: 09425143654

Editorial

The ESW V[th] Annual National Research Conference on **"Sustainable Development of Ecosystem, Wildlife and Heritage Conservation for Human Welfare 30 and 31 January, 2018 organized by Environment and Social Welfare Society (ESW Society), Khajuraho** has its inception when The Official Agenda for Sustainable Development adopted on 25 September 2015 for Sustainable Development Goals and its associated 169 targets. as the thrusty area for work in Climate action provide a field for research and discussion.

Since the first United Nations Conference on Environment and Development in 1992 - known as the Earth Summit, it was recognized that achieving sustainable development would require the active participation of all sectors of society and all types of people. Agenda 21, adopted at the Earth Summit, drew upon this sentiment and formalized nine sectors of society as the main channels through which broad participation would be facilitated in UN activities related to sustainable development. These are officially called "Major Groups" and include the following sectors:

- ☆ Women
- ☆ Children and Youth
- ☆ Indigenous Peoples
- ☆ Non-Governmental Organizations
- ☆ Local Authorities
- ☆ Workers and Trade Unions
- ☆ Business and Industry
- ☆ Scientific and Technological Community
- ☆ Farmers

Two decades after the Earth Summit, the importance of effectively engaging these nine sectors of society was reaffirmed by the Rio+20 Conference. Its outcome document "The Future We Want" highlights the role that Major Groups can play in pursuing sustainable societies for future generations. In addition, governments invited other stakeholders, including local communities, volunteer groups and foundations, migrants and families, as well as older persons and persons with disabilities, to participate in UN processes related to sustainable development, which can be done through close collaboration with the Major Groups. Major Groups and other stakeholders (MGoS) continue to demonstrate a high level of engagement with intergovernmental processes at the UN. The coordination of their input to intergovernmental processes on sustainable development has been led by UNDESA/ Division for Sustainable Development (DSD). Member States ultimately decide upon the modalities of participation of MGoS. Thus, the engagement and participation of MGoS in intergovernmental processes related to sustainable development varies depending on the particular sustainable development topic under discussion.

Since the 1st ESW National conference on sustainable development of natural resources and wildlife conservation, convened by ESW Society in Khajuraho, Madhya Pradesh India in 2014, a growing body of knowledge has been generated addressing the complex relationships between the Nature conservation and wildlife with important research activities on this subject. There is now a wide recognition of the urgent need for the environment, biodiversity, and tourism industry, national governments and international organizations to develop and implement strategies to face the global warming and to take preventive actions for future effects, as well as to mitigate tourism's environmental impacts contributing to global warming. Furthermore, such strategies should take also into account the needs of developing countries in terms of Millennium Development Goals.

The **Millennium Development Goals (MDGs)** were the eight international development goals for the year 2015 that had been established following the Millennium Summit of the United Nations in 2000, following the adoption of the United Nations Millennium Declaration. All 189 United Nations member states at that time, and at least 22 international organizations, committed to help achieve the Millennium Development Goals by 2015. Each goal had specific targets, and dates for achieving those targets. Although there has been major advancements and improvements achieving some of the MDGs even before the deadline of 2015, the progress has been uneven between the countries. In 2012 the UN Secretary-General established the "UN System Task Team on the Post-2015 UN Development Agenda", bringing together more than 60 UN agencies and international organizations to focus and work on sustainable development. At the MDG Summit, UN Member States discussed the Post-2015 Development Agenda and initiated a process of consultations. Civil society organizations also engaged in the post-2015 process, along with academia and other research institutions, including think tanks.

The **Sustainable Development Goals (SDGs)**, officially known as **Transforming our world: the 2030 Agenda for Sustainable Development** is a set of seventeen aspirational "Global Goals" with 169 targets between them. Spearheaded by the United Nations, through a deliberative process involving its 193 Member States, as well as global civil society, the goals are contained in paragraph 54 United Nations Resolution A/RES/70/1 of 25 September 2015. The Official Agenda for Sustainable

Development adopted on 25 September 2015 has 92 paragraphs, with the main paragraph (51) outlining the 17 Sustainable Development Goals and its associated 169 targets.

Keeping above serious issue in mind ESW Society, India President Dr. Ashwani Kumar Dubey has called for action on Quality Education; Clean Water and Sanitation; Climate Action; Life on Land; Peace, Justice and Strong Institutions; Partnerships for the Goals, and Nature conservation to be taken in close coordination with global action on The *Transforming our world: the 2030 Agenda for Sustainable Development.* To provide a platform to Educational Administrators, College Principals, Deans, Readers, Head of Departments, Professors, Assistant Professors, Scientists, Environmentalist, Stakeholders, Researchers, Young scientists and Students to disseminate knowledge related to Nature Conservation, Resource Management and possible solution by Technological Approach.

Dr. Ashwani Kumar Dubey

Contents

About ESW GAST IJGSR

Environment and Social Welfare Society (ESW Society) *Dedicated to Environment, Education and Sciences and Technology entire India since bi-Millennium* is an ISO 9001:2015 certified organization of the India. Now it's worldwide known by its impact. ESW Society has been to develop relationship between Environment and Society envisions the promotion of Education and Sciences among the University, College and School students as well as in the society for Environment and Social welfare as well as Human Welfare.

It is registered under the society Act 1973, Government of Madhya Pradesh, India on 31 January 2000 with No SC2707. It was affiliated by Nehru Yuva Kendra Sangathan, Ministry of Youth Affairs and Sports, Government of India. It accredited by Madhya Pradesh Jan Abhiyan Parishad, Government of Madhya Pradesh, since 2013, also enrolled in Navankur Yojana with enrollment number NV2016CHH0062 Dated 29/09/2016. It is also registered with NGO-PS, Government of India And having The NGO-Partnership System, Portal (NGO-DARPAN), National Institution for Transforming India (NITI) Aayog, Govt. of India. ID MP/2014/0076324. NGO Databases.

Godavari Academy of Science and Technology (GAST): GAST *Dedicated to Environmental Sciences and Bio-technology.* The Academy was established as prompt Science Academy in India. ESWSociety establish Godavari Academy of Science and Technology in 2010. Goal of Academy is understand the problem, undertaking and solve them by scientific approach of Scientist or Researcher Organizing Scientific Session, Science Communication Activity, Seminar, Symposium, Conferences, Workshop, Popular Science Lecture, Interaction with leading scientist, Quiz Contest, Interactive Session on Health and Pollution Control and Career in Biological Sciences are main program aim to develop Scientific temper among students and Social workers.

Co-operating with other institute in State and in India as well as Abroad, having similar objects and to appoint coordinators of the Academy to act National and International Bodies. **Website**: www.godavariacademy.com

International Journal of Global Science Research (IJGSR): IJGSR is Open access, Peer-reviewed, Refereed, Indexed, Biannual, Bilingual, Impacted Academic Journal

Abbreviated title: Int. J. Glob. Sci. Res.

ISSN (online): 2348-8344 Paris

Journal Website: www.ijgsr.com

Started Year *April 2014;*

Subject: Environmental Sciences

Frequency: Bi-annual (April and October)

Language: English and Hindi

Index Copernicus International ICV 2016: 62.6

NAAS score 2018: 2.76

Crossref DOI: 10.26540

Impact Factor 2017: 2.312

License: Free for non-commercial use

Email: editor@ijgsr.com

Keywords: Global Science, Environmental Sciences, Biosciences, Earth and Atmospheric Sciences

Under auspicious of: Environment and Social Welfare Society, India

Go to Journal: www.ijgsr.com

Brief Report of

ESW V[th] Annual National Research Conference on "Sustainable Development of Ecosystem, Wildlife and Heritage Conservation for Human Welfare"

Organized By: Environment and Social Welfare Society Khajuraho-471606 MP, India.

In association with: The National Academy of Sciences India, Allahabad, UP and Maharaja Chhatrasal Bundelkhand University, Chhatarpur MP

Assisted by: Godavari Academy of Science and Technology, Chhatarpur, Madhya Pradesh on **30 and 31 January, 2018**

Website: www.godavariacademy.com and www.ijgsr.com

Prof. P. K. Verma, Vice-Chancellor Inaugurating ESW 5th National Conference.

A PRELUDE: After the success of 4rd National conference on "Strategy for Human Welfare on Nature conservation and Resource management" The **IV ESW National conference on "Impact of Global warming on Environment, Biodiversity and Ecotourism"** Organized By ESW Society Khajuraho-471606 Madhya Pradesh, India during 30 and 31 January, 2017 at UNESCO World Heritage Khajuraho, India.

Object: To provide a platform to Vice Chancellors, Educational Administrators, College Principals, Deans, Professors, Readers, Associate Professors, Assistant Professors, Scientists, Environmentalist, Researchers, Young scientists and Post Graduate Students to disseminate knowledge related to **Ecosystem, Wildlife National and World Heritage.**

Theme: To take some positive steps towards improving our Earth for future generation.

Goal: The moral obligation to act sustainably as an obligation to protect the Natural processes that form the context of human life and culture, emphasizing those large biotic and abiotic systems essential to human life, health, and flourishing culture. Ecosystems, which are understood as dynamic, self-organizing systems humans have evolved within, must remain 'healthy' if humans are to thrive. The ecological approach to sustainability therefore sets the protection of dynamic, creative systems in Nature as its primary goal. The principal goal of this conference will be to present some of the latest outstanding breakthroughs in Environment and global health, to bring together both young and experienced scientists from all regions of the world, and to open up avenues for research collaborations at regional and global level.

INAUGURAL FUNCTION: The **V ESW National Conference inaugurated on 30 January, 2018 by Chief Guest** Prof. P. K. Verma, Vice Chancellor, Barkatullah University Bhopal, Madhya Pradesh. Key note speaker Prof. Prakash Singh Bisen, Former Vice Chancellor, Jiwaji University, Gwalior, Madhya Pradesh, Prof. K. K. Sharma, Former Vice Chancellor, MDS University, Azmer, Rajasthan, President Prof Bhartendu Prakash, Chhatarpur and Fellow/Member of Environment and Social Welfare Society Khajuraho, India, Mrs. Vandana Dubey, Managing Director, Godavari Academy of Science and Technology, Chhatarpur, MP and other distinguished guests, participations from various part of India and One hundred fifty + listener including media were participated in conference.

Souvenir released with Message of Honourable Om Prakash Kohli, Governor of Madhya Pradesh, Dr. Kailash Chandra, Director, ZSI, Ministry of Environment and Forest, Govt. of India, Kolkata; Prof. P. K. Verma Vice Chancellor, Barkatullah University, Bhopal, MP and Prof. Priyvrat Shukla, Vice Chancellor Maharaja Chhatrasal Bundelkhand University, Chhatarpur. Abstract with Eighty five Research Abstract which included from various State of India *viz.* Jammu and Kashmir, Himachal Pradesh, Madhya Pradesh, Uttar Pradesh, Chhattisgarh, New Delhi, Bihar, Maharashtra, Rajasthan, Odisha, Gujarat, Uttarakhand, Punjab, West Bengal, Chennai. And from abroad Romania and Syria.

Released Souvenir by Guest.

Prof. P. K. Verma, Vice Chancellor, Barkatullah University Bhopal, Madhya Pradesh addressed on World heritage conservation.

Prof. Prakash Singh Bisen, Former Vice Chancellor, Jiwaji University, Gwalior **address Key note on Food Heritage.**

Prof. K. K. Sharma, Former Vice Chancellor, MDS University, Azmer, Rajasthan highlighted on Myths, facts and new technologies related to Snake Bite Free India.

Prof. Bhartendu Prakash, Chhatarpur focussed on Bundelkhand development, livelihood and River conservation.

Dr. Ashwani Kumar Dubey, Executive Director, ESW Society and President and Organizing Secretary of The National conference delivered his presidential address emphasized the role of ESW Society in Ecosystem, Wildlife and Heritage conservation and also focus on annual report of the ESW Society, Khajuraho.

TECHNICAL SESSION: After the inauguration, the Technical session held Twenty two Research paper and Six poster presented in the two days technical session.

The general topics covered in the conference will be as under Ecosystem, Wildlife and Heritage conservation and Technological Approach Lab to Land.

SCIENTIFIC EXIBITION: An exhibition was arranged along with conference. Researchers got opportunity with delegates and scientist to discuss their needs and publication in the reputed journals.

CULTURAL PROGRAMME: To conserve, promote and develop the Indian's culture, ESW Society arranged cultural event with the national conference.

Participants of ESW Vth Annual National Research Conference.

CELEBRATION OF 18th FOUNDATION DAY OF ESW and AWARD CEREMONY ON 31 JANUARY: Dr. K. S. Tiwari, Former Director, IGNOU, Bhopal MP was the Chief Guest; Prof. K. K. Sharma, Former Vice Chancellor, Maharishi Dayanand Saraswati University, Ajmer, Rajasthan, Prof Prakash Singh Bisen, Former Vice Chancellor, Jiwajii University, Gwalior, MP and Prof Anand Kar Department of Life Sciences, Devi Ahilya Bai University, Indore MP were Special Guest and Dr. Shivesh Pratap Singh, Secretary BER Chapter, NASI, Chitrakoot was the President of the 18th Foundation Day of ESW and award ceremony of the Conference. And other eminent scientists were present on this occasion.

Lifetime Achicvement Award: Prof Anand Kar, Department of Life Sciences, Devi Ahilya Bai University, Indore M.P.

Best Paper Oral Presentation Award in each Session awarded to Pooja Chahar, Madhya Pradesh, Mr. Sajjad-ul-Akbar-Wani, Jammu and Kashmir, Dr. Ravikant Bhatiya, Himachal Pradesh and Mr. Arjun Shukla, Madhya Pradesh.

Best Poster Presentation Award in each session awarded to Dr. Smita Singh and Dr. Rekha Sharma, Madhya Pradesh

Young Scientist Award (Below 35 Years) to Dr. Ravikant Bhatiya, Himachal Pradesh.

Young Environmentalist Award: Mr. Sajjad-ul-Akbar-Wani, Jammu and Kashmir.

Godavari Academy Impact Award to Dr. K. S. Tiwari, Former Director, IGNOU, Bhopal, MP.

**National Amazing Godavari Memorial Award (NAGMA) in the Field of
Education and Science Awarded to Prof. Pramod K. Verma, Vice-Chancellor,
Barkatullah University, Bhopal, MP.**

ESW Recognition Award: Dr. Atul Kumar Mishra, Kanpur, UP and Dr. Ganpathi Venketsubramaniyam, Anna University, Chennai.

Fellowship of ESW Society Awarded to **Mrs. Shivani Rai**, Madhya Pradesh, Dr. Hemlata Pant, Uttar Pradesh and Dr. Esha Yadav, Uttar Pradesh

Certificate of Paper presenter and Participants given by the Chief Guest.

And Mementos presented by Mrs. Vandana Dubey MD Godavari Academy of Science and Technology to our Guest.

Vote of thanks by Dr. Ashwani Kumar Dubey, National President, ESW Society, Khajuraho

Recommendations

☆ Khajuraho is a confluence of Ecosystem, Wildlife and Heritage. Sustainable developments of these assists are necessary for Tourism development.

☆ Bundeli banquet are suitable according to Bundelkhand climate We should take bundelkhand food promptly. We should conserve and protect its trees also at local level. These Banquet implimated in Hotels also.

☆ We should protect Amphibians such as Frog and Reptiles viz Lizards and Snakes to protect and balance our Ecosystem.

☆ Mountains, Rivers, Lakes and aquatic ecosystem are all necessary to keep our balanced environment. There is an urgent need to think over these environmental hazards by both State and Central Government.

☆ Fauna and Flora of India may attractive for the pilgrims. There are several ancient temples related with God and Goddess.

☆ Ancient ruined Temple, Stepwell, Palace, Garhi, Tomb, Monument related with old Kings and Queens of different dynasty, accessibility should be developed to reach over there.

210+ participants were present out of these General 62 per cent Schedule Caste 12 per cent Schedule Tribes 25 per cent and Women more than 30 per cent overall.

News Gallery

National news paper, Local news paper and National, State and Local electronic channel covered this event promptly.

Chapter 1

Role of Wildlife in Human Welfare with Special Reference to their Medicinal Values: A Strategy for Conservation

Anand Kar

Professor and Chairman Board of Studies in Life Sciences,
School of Life Sciences, Devi Ahilya Vishwavidyalaya, Khandwa road,
Indore – 452 017, India
E-mail: karlife@rediffmail.com

According to Cambridge English dictionary, wildlife means "animals and plants that grow independently of people, usually in natural conditions". In fact, any form of life away from natural habitat, that is not domesticated (in case of animals) or cultivated (in case of plants) can be termed as wildlife. Many wild plants and animals look apparently not so important. However, directly or indirectly all play great role in the human welfare. However, because of different reasons/factors such as illegal felling of trees, deforestation and natural disasters such as flood, drought, earth quack, tsunami, oil spill and forest fire their number is decreasing every year. Here I give an outline on their importance with particular reference to their medicinal values.

Our investigations indicate that all most all plants except the weeds directly or indirectly contribute in the welfare of human being. The importance of many plants as source of food and fodder is well known. However, they are now known for their different values including ecological, economical, aesthetical, agricultural, and scientific values. One of the most important present day contributions of wild plants is in the development of drugs/medicines. It is believed that about 80 per cent of global village population depends on plant based drugs. Some plants even have the potential to regulate important present day problems such as thyroid disorders, diabetes mellitus and cardiovascular problems (Kar and Panda, 2004; Panda *et al.*, 2009, 2017). Similarly, some wild animals not only have ecological and recreational values, but too are known to have medicinal values. Obviously, their conservation and propagation is very much required. Therefore, a strategy for their management has been suggested. This includes prevention of deforestation, poaching/killing of animals, provision of more forest staff and fund, stringent wildlife act, scientific breeding and innovative research. All need to put efforts for their management.

REFERENCES

1. Kar, A and Panda S (2004).*Scientific Basis of Ayurvedic Therapies"*. L. Mishra,Ed. Chap. 8, CRC press, USA.

2. Panda S, Biswas S and Kar A (2017). *Scientific-Reports*-Nature.com, 7: 16146.

3. Panda S, Jafri M and Kar A, and Meheta BK (2009) *Fitoterapia*, 80(2): 123-126.

Chapter 2

Heritage and Tourism

Mangla Sood

Department of Zoology,
Govt. Girls P.G. College of Excellence,
Sagar, Madhya Pradesh

Tourism, as a form of community revitalization is booming in today's time. Importantly, in the last several decades, along with its scale, the natureof tourism has also changed. As social and technological changes made tourism more affordable and accessible for millions of people, theonce-traditional and long-awaited family summer vacation to the shorebecame just one option among many that beckon all year round. This shift in tourism from relaxation to self-discovery is reflected in the explosion of niche market designations within the tourism industry. The more widely known include adventure tourism, culinary tourism, religious tourism, ecotourism, sustainable tourism, and educational tourism. Cultural heritage tourism is one of the fastest growing specialty markets in the industry today. This paper attempts to analyze different aspects of heritage and tourism.

Introduction

An unprecedented number of people are travelling around the world, and the figures are only expected to rise, with international arrivals growing from 25 million in the 1950s to 1.2 billion in 2016 and to 1.8 billion by 2030. Having outperformed the global economy for the sixth consecutive year, the industry has proven to be resilient to both geopolitical uncertainty and economic volatility. The aviation, travel and tourism industry accounts for 10 per cent of global GDP and 10 per cent of jobs on the planet. The industry should be a priority for countries around the world given its ability to make a real difference to the lives of people by driving growth, creating jobs, reducing poverty and fostering development and tolerance.In recent years, India has tremendously improved its travel and tourism competitiveness, rising from 65th position in 2013 to 40th position in The Travel and Tourism Competitiveness Report 2017. India is now one of the fastest-growing aviation markets in the world, with its domestic demand reaching nearly 100 million passengers. Yet international arrivals have remained relatively low, at 9 million, providing India with a unique opportunity to consider how to build demand and create adequate supply for its travel and tourism industry. Focusing on its opportunities and understanding its current limitations will allow India to realize its objective of welcoming over 15 million foreign tourists by 2025 and becoming the largest aviation market by 2030. Already endowed with incredible natural beauty and a unique cultural heritage and diversity, India must enhance its value proposition and foster an enabling environment for the industry to prosper.

Cultural Heritage Tourism: Concept and Significance

The curiosity to know about the others, understand the nuances of their culture and heritage and appreciate the differences, are the main reasons for the growth and development of cultural-heritage tourism. Technological and communication developments today have made possible for man-kind to see, read and hear about the life-styles ofdifferent places or countries, more than ever before. Cultural-heritage tourism is relatively a new concept in the system ofmodern tourism. It emerged in the 1970's and was recognized by UNESCO in the year 1976. Some scholars feel that even though cultural-heritage tourism is a relatively new concept, yet it is not a new phenomenon in itself. Carrol Van West is of the opinion that cultural-heritage tourism did not spring up as a new concept in the last decade; rather, there already existed three main types of cultural-heritage tourism projects. Three projects being: 1. The Williamsburg's living history type, 2. The old town recreation with buildings brought in form of a variety of locations and restored to create a new sort ofhistory theme-park towns and 3. The retention and preservation of existing structures in a town, by using the historical district ordinances used in New Orleans to maintain that city's historical character.

Distinct Characteristics of Cultural Heritage Tourism

1. Cultural-heritage tourism is a delightful special interest activity of travelling
2. Curiosity and thirst for knowledge are crucial in cultural-heritage tourism, travelling without which can never be accounted for cultural tourism.
3. Not only past, but contemporary lifestyle *etc.*, are also important components of cultural-heritage tourism.
4. Cultural-heritage tourism is a deliberate, serious and systematic cognitive pursuit.
5. In cultural-heritage tourism, visit to a tourist destination attraction is not en passant, the visit is always purposeful
6. There is an extra pull in cultural-heritage tourism to entice tourists to make the most of their leisure time.
7. Cultural-heritage tourism is a measure to learn and spread ideas, thoughts and facts about different cultures with sagacity.
8. Cultural-heritage tourism is connected with articles offaith and places and common or occasional occurrences

Principles Governing Cultural Tourism

The International Cultural Tourism Charter (8th Draft ICOMOS, 12th General Assembly, Mexico: October, 1999) has outlined principles governing international cultural tourism, some of which are reproduced below to provide a valuable insight into the various aspects of cultural-heritage tourism.

1. The programmes for the promotions and conservation of the physical attributes, intangible aspects, contemporary cultural expressions and broad context, should facilitate an understanding and appreciation ofthe significance of heritage by the host community and the visitor in an equitable and affordable manner.

2. Individual aspects of natural and cultural heritage have differing levels of significance, some with universal values, others of national, regional or local importance. Interpretation programmes should present that significance in a relevant and accessible manner to the host community and the visitor, with appropriate, stimulating and personal explanation of historical, environmental and cultural information.

3. Interpretation and presentation programmes should facilitate and encourage the high level of public awareness and necessary support for the long-term survival ofthe natural and cultural heritage.

4. Interpretation programmes should present the significance of heritage places, traditions and cultural practices within the past experience and present diversities of the area and the host community, including that of minority 102 cultural or linguistic groups. The visitor should always be informed of the differing cultural values that may be ascribed to a particular heritage resource.

5. The long term protection and conservation ofliving cultures, heritage, places, collections, their physical and ecological integrity and their environmental context, should be an essential component of social, economic, political, legislative, cultural and tourism development policies.

6. Tourism projects, activities and developments should achieve positive outcome and minimize adverse impacts on the heritage and lifestyle of the host community, while responding to the needs and aspirations of the visitor.

7. Conservation, interpretation and tourism development programmes should be based on a comprehensive understanding ofthe specific, but often complex or conflicting aspects of heritage significance of the particular place.

8. The retention of the authenticity of heritage, places and collections is important. It is an essential element oftheir cultural significance, as expressed in the physical material, collected memory and intangible traditions that remain from the past. Programmes should present and interpret the authenticity of places and cultural experiences to enhance the appreciation and understanding of that cultural heritage.

9. Tourism development and infrastructure projects should take account of the aesthetic, social and cultural dimensions, natural and cultural landscape, biodiversity, characteristics and the broader visual context of heritage places. Preference should be given to using local materials and take account of local architectural styles or vernacular traditions.

10. Before heritage places are promoted or developed for increased tourism, management plans should assess the natural and cultural values of the resources. They should then establish appropriate limits of acceptable change, particularly in relation to the impact of visitor numbers on the physical characteristic, integrity, ecology and biodiversity ofthe place, local access and transport systems and the social, economic and cultural well being ofthe host 103 community. If the likely level of change is unacceptable the development proposal should be modified.

Cultural Heritage Tourism Management

Application of management techniques to link cultural-heritage and tourism and to obtain the desired results of cultural-heritage tourism is 'cultural-heritage tourism management'. In simple words, those processes, functions and activities, which establish a link between culture and tourism to achieve the desired objectives of satisfying the cultural curiosity of a tourist with interest in present and past lifestyles, places, objects and activities of the host country, region or community together constitute Cultural-Heritage Tourism Management. It deals with setting, seeking and reaching objectives of cultural-heritage tourism through the application of right management practices with effective use ofnatural and human resources. Cultural-Heritage Tourism Management cannot emerge without integrating all aspects of management with basic components of culture and tourism. All essential elements and forms of culture and heritage, appropriately interlinked with the essential elements and forms of tourism together with the application of right kind of management practices to satisfy the aspiration and demands of the tourists besides 106 taking care of the interests and needs of the host community together constitute the basic tenets of Cultural-Heritage Tourism Management. Cultural-Heritage Tourism Management can play an important role in the development of better understanding between the nations ofthe world. In the Manila Declaration of World Tourism, 1980, it was emphasized that, "Cultural-Heritage Tourism can be a vital force for world peace and can provide the moral and intellectual basis for international understanding and independence" (Manila Conference Report 1980).

Cultural Heritage Tourism in India

India - much beyond the perception ofbeing mystic land ofsnake charmers - offers a dazzling array of destinations and experiences in the realm of cultural heritage. It is a 109 huge country of crowded millions comprising one-fifth of the world-population, making India the seventh largest country in the world occupying 2.3 per cent of the earth's land surface, stretching across 3000 Kms. approximately east to west and north to south. Bounded in the north by the majestic Himalayan ranges to the spectacular coastline in the south washed by three seas, India is a vivid kaleidoscope of lucid landscapes, splendorous historical and architectural monuments, archaeological wonders, golden beaches, colorful people, fairs and festivities. Today, it is a land relatively untouched by the vagaries of modem developed world and lies largely unexplored by travelers from around the globe. Yet, it is ironical that not long ago, its glory, fame, cultural heritage and wealth attracted invaders from all over the world. From Alexander onwards, the Parthians, the Kushanas, the Turks, the Mughals, the Portuguese, the French the Dutch and the British, all ofthem came to this country, captured our lands but could not conquer the minds or alter the cultural fabric and social ethos ofthe people ofthis country. Why is it then, very few people visiting India, the land of one billion people, despite its rich cultural heritage of ancient civilization? The figure of 2.6 million people comprising 0.38 per cent of the international travelers, visiting this country has become stagnant in the last one decade. In fact, what is not known to a large number of people is that 3.8 million visitors go abroad every year, which far exceeds than visitors to this country. Foreign tourist arrivals to India, ifincreased tenfold to even 3.8 per cent of the international tourist arrivals, which is equivalent to 27 million people visiting China every year, can virtually help in wiping out the entire budgetary deficit ofthe country besides financing the entire budget ofthe Government ofIndia. Alas! The fact remains that this is a far cry, since even the target of achieving one per cent ofinternational tourist arrivals is nowhere in sight. Despite tremendous potential and cascading effect of tourism on economic growth and community development, is it not surprising that the first ten years of India's Five-year Plans did not even mention tourism among the sectors of economic priority? Is it because of the misplaced priorities and general apathy of the Indian policy makers and agencies that even today, tourism has had no patron even in the Constitution ofIndia, where it is neither listed as a central nor as a state or concurring subject? According to World Travel and Tourism Council (WTTC) reports India 110 ranks 153rd out of 160 countries in terms of government expenditure on tourism. An analysis of Five-Year Plans reveals that tourism accounts for less than 0.2 per cent of the total plan layout ofGovernment ofIndia. Even at the conceptual level, there is no bigger dilemma in government where both at the central and slate government levels (with few exceptions like Rajasthan), tourism and culture are separate departments, chalking out their individual pathways on to old beaten tracks and policies eventually leading to a situation where the left hand does not know what the right hand is doing. The strategy so far of developing tourism, not to speak of cultural-heritage tourism, if any, has been like running around in a circle, which has no beginning or end. The real essence ofIndian culture as reflected in our lifestyles, customs, traditions, performing arts, spiritualism, glories of the past, historical monuments, archaeological sites and ancient ruins, are what actually fascinates and attracts visitors from all over the world. Unfortunately, this essence of linking culture with tourism presenting India - that really is, has always been confined to fringes with marginal importance.

Archaeo Heritage Tourism

In the context of a cultural-heritage tourism market, Archaeo-Heritage tourism can be described as visiting archaeological sites and landscapes—the material remains left by past cultures. Travel and viewing are acts of cultural understanding that attribute meanings and values to sites. It is widely recognized that a knowledge and understanding of the origins and development of human societies is of fundamental importance to humanity in identifying its cultural and social roots. The archaeological heritage constitutes the basic record of past human activities. Archaeological heritage can be seen as an asset, both, in terms of representational value and as a source of revenue. Indeed, tourism can be the most effective vehicle for archaeology's ethical obligation to public education and outreach. Similarly, through tourist's interest in archaeological sites and the income it generates, local communities can appreciate the symbolic and economic values of archaeological sites and participate in their protection and conservation.

Why Archaeo Heritage Tourism

1. Archaeo-heritage tourism has positive economic and social impact on country, region and destination.
2. Archaeological Heritage and Cultural features are essential in building a country's image. It establishes and reinforces identity of a country, region and destination.
3. It helps to preserve the cultural and archaeological heritage.
4. With culture as an instrument it facilitates harmony and understanding among people.
5. Archaeo-heritage tourism supports culture and helps renew tourism

Recommendations for Promoting Cultural Heritage Tourism

1. Take advantage of 600,000 villages with their own cultures and heritage, ecotourism and cruise tourism to create unique experiences for travelers.
2. Integrate the Incredible India campaign into a holistic campaign that includes not only print but also other channels, such as digital, social, placement, review sites and global media – and that focuses on the positives of visitor-created content, while addressing the challenges these visitors report
3. Enhance the perception and reality of India as a safe destination by designing and implementing enhanced security protocols
4. Invest in both physical and digital infrastructure development to confront the issue of last mile connectivity, hazardous road travel and the lack of affordable hotels hampering international travelers' experiences, while elevated taxes hinder the industry's profitability
5. Take advantage of the labor force available in India to provide a quality product to tourists, by training skilled and unskilled workers in the hospitality industry through both public and private programmes.

Conclusion

To reach its goal of becoming the world's largest aviation market by 2030 and welcoming over 15 million international tourists by 2025, India must enhance its value proposition and foster a conducive environment. By doing so, it will be better able to build demand and attract tourists, while simultaneously creating adequate supply through the development of physical and digital infrastructure and progressive legislation.India has everything – from over 7,000 km of coastline, and rain forests, deserts and snow-capped mountains, to temples, mosques, wildlife, tribal habitation and a multicultural population. Yet it has neglected its travel and tourism industry's potential. With so many attributes, India's challenge is not to build sites but rather experiences around what it already has

REFERENCES

1. Organisation for Economic Co-operation and Development (OECD) (2017), Government at a Glance 2017, OECD Publishing: Paris.

2. J. Walter Thompson Intelligence (2013), "Study: Constantly connected Millennials crave sensory experiences", 25 January 2013.

3. Homi Kharas (2017), "The Unprecedented Expansion of the Global Middle Class", Global Economy and Development Working Paper 100, the Brookings Institution.

4. World Travel and Tourism Council (2017), Travel and Tourism Economic Impact 2017, WTTC: London.

5. World Tourism Organization (UNWTO) (2011), "Tourism towards 2030: Global overview", UNWTO General Assembly 19th Session, Gyeongju, Republic of Korea, 10 October 2011.

6. World Tourism Organization (UNWTO) (2016), Visa Openness Report 2015, UNWTO: Madrid.

7. World Travel and Tourism Council (2017), "India is the world's 7th largest tourism economy in terms of GDP", Press Release, 4 April 2017.

8. CAPA Centre of Aviation (2016), "CAPA India Aviation Outlook 2017/18: Surging traffic but infrastructure constraints become critical", 30 December 2016.

9. The Guardian (2017), "Barcelona cracks down on tourist numbers with accommodation law", 27 January 2017. 10 World Tourism Organization (UNWTO) (2017), Tourism Highlights, 2017 Edition, UNWTO: Madrid.

Chapter 3

Effects Studies of EDC in Silver Carp Fish of Upper Lake, Bhopal with Reference to Vitelogenin Level

Kalpana Vishwakarma[1]*, M. Murlidhar[1] and Ruchira Chaudhary[2]

[1]*Chanakya College, Bhopal, Madhya Pradesh*
[2]*Sarojni Naidu Govt. Girls P.G. (Auto) College, Bhopal, Madhya Pradesh*
*E-mail: vkalpana855@gmail.com

Bhojtal or Upper Lake of Bhopal has been built by a king of Malwa Parmar Raja Bhoj in his tenure (1005-1055). Bhopal City has grown around this Lake. This Lake is a major source of drinking water and serving around 40 per cent of resident of this City. This is also used for agriculture and fishing purpose. It has 31km area and has 361km total catchment or watershed area. This Lake is finally drained into Kaliasot River. Level of Vitellogenin in serum of Silver carp was recorded in different season to study endocrine disturbance of this fish. Vitellogenin is a yolk precursor protein, who contributes oocyte development and become go for fertilization and then embryo develops. Level of Vitellonenin was determined by using ELISA kit method with standard of Silver carp vitellogenin. Reported normal range of vitellogenin in Silver carp fish is 7.5ng/ml to 500ng/ml in female. This study reported 398.67±43.28 ng/ml in February 2016, 1586.67±288.33 ng/ml (this is due to before breeding season) in May 2016, extensively it was reported 268±23.93 ng/ml in November, 2016. Reported level of vitellogenin in serum of female Silver carp is almost near to normal range. This means Upper Lake of Bhopal is almost good to use for drinking purpose and is a non polluted Lake or has less than permissible limit of pollution. This is needed to look unfair activities in this Lake for their long time survival.

Keywords: Oogenesis, Vitellonenin, Upper Lake, ELISA, Breeding.

Introduction

Bhojtal or Upper Lake of Bhopal has been built by a king of Malwa Parmar Raja Bhoj in his tenure (1005-1055). Bhopal City has grown around this Lake. This Lake is a major source of drinking water and serving around 40 per cent of resident of this City. This is also used for agriculture and fishing purpose. It has 31sq.km area and has 361sq.km total catchment or watershed area. This Lake is finally drained into Kaliasot River (Dwivedi and Choubey, 2008). Silver carp (*Hypophyhalmichthys molitrix*) is a variety of Asian carp and member of *Cyprinidae* family. Colour of this fish is silver with darker colour on back (Robison and Buchanan 1988). Average length of this fish is 1meter and 27kg in weight. It tolerates about 12ppt and low DO (3mg/L). It eats both phytoplankton and zooplankton feed (Radke and Kahl, 2002). This is an eaten fish (US Geological Survey, 2018). Endocrine disrupting chemicals found in various pesticides, contaminated foods, heavy metals, personal care products, paints, beauty products, baby care products and others. City sewage and garbage has lot of EDC which finally goes to water bodies. EDCs have been alleged to be linked with altered reproductive function in male as well as in female. It increases risk of breast cancer, abnormal growth in embryo, changes in immune function *etc.* Pregnant mother are more susceptible to EDC (WHO, 2018). Vitellogenin is a biomarker for endocrine disrupter. This is a yolk precursor protein, who contributes oocyte development and become go for fertilization and then embryo develops. Level of Vitellogenin was determined through ELISA method (Hansen *et al.*, 1998). EDC quantification, physiological study and vitellogenin level was investigated in this study for effect study of water quality and EDC on fish health.

Materials and Methods

pH, TDS, temp and DO of sampled water were analyzed according to APHA, 1998. Pb and Cd was determined according to Agemian and Chau, 1975. DDT was estimated as per Johnson and Finley, 1980. Physiological status was investigated as Wanner and Klumb, 2009 methods. ELISA was performed for vitellogenin test with Biosence Laboratory Kit method.

Results and Discussion

Pollution of water body is determined according to investigation report and in comparison to standard values derived to fit for drinking of water by WHO. This investigation revealed pH 7.63±0.12, 7.93±0.07, 7.70±0.09; 173.67±10.34, 184.67±4.50, 177.67±5.79; temperature 19.33±1.25, 33.67±1.25, 18.33±1.25; DO 7.17±0.33, 6.47±0.49, 6.87±0.29 in Feb, 2016, May, 2016, Nov, 2016 respectively. Earlier report of Singh and Srivastava, 2016 reported 6.72 to 8.12 in different station of Upper Lake. This pH is in between range of BIS standard. Temperature in °C denoted 18.33±1.25 in winter and 33.67±1.25 in summer is good for survival of aquatic lives in this Lake. Mosely, 1983 express that temperature of water is varies due to atmosphere temperature in mesotrophic water bodies. TDS and DO revealed limit to fit for drinking and aquatic life. Higher TDS is an indicator of pollution (Tay, 2007) and increased organic matter in water (Phiri *et al.*, 2005). Good DO reported in every season suggest that there are very less pollutant (Srivastava *et al.*, 2009). Investigation of EDC *viz.*, DDT, Pb and Cd were performed and reported 0.007±0.001 mg/L concentration of Pb only in summer. This told the story of upper Lake is a non polluted water body (Anu, 2011; Bajpai, 2009). Table 3.3 presented physiological conditions and level of vitellogenin in Silver carp of Upper Lake in Feb, 2016, May, 2016 and November,

2016. Length, weight and condition index report is near idealist values of this fish and support the study of Wanner and Klumb, 2009; Papoulias, 2006. GSI (Gonado somatic index), HSI (Hepato somatic index) are gradually increased during breeding season (Jacques and Patric, 2003). Increased GSI and HSI reported in this study reveals breeding season, but in normal range (Sabrina *et al.*, 2016). Vitellogenin level Silver carp was investigated and reported in Table 3.3 tell the increased value found in breeding season and in normal range (Sumpter and Jobling, 1995). There are no any burden of pollution in this fish as recorded their vitellogenin level (Burzawa and Dumas, 1991).

Table 3.1: Water Parameter Status of Upper Lake in different Season

Water Parameter	February, 2016	May, 2016	November, 2016
		pH	
Mean±SD	7.63±0.12	7.93±0.07	7.70±0.09
		TDS (mg/L)	
Mean±SD	173.67±10.34	184.67±4.50	177.67±5.79
		Temp	
Mean±SD	19.33±1.25	33.67±1.25	18.33±1.25
		DO (mg/L)	
Mean±SD	7.17±0.33	6.47±0.49	6.87±0.29

Table 3.2: EDC Concentration in Upper Lake in different Season

EDC	February, 2016	May, 2016	November, 2016
		DDT (mg/L)	
Mean±SD	0	0	0
		Pb (mg/L)	
Mean±SD	0	0.007±0.001	0
		Cd (mg/L)	
Mean±SD	0	0	0

Table 3.3: Physiological Parameters and Vitellogenin Level in Silver Carp of Upper Lake

Parameters	February, 2016	May, 2016	November, 2016
		Length (In inch)	
Mean±SD	14.03±0.09	14.07±0.12	14.07±0.12
		Total Body Weight (In Kg)	
Mean±SD	3.57±0.26	3.7±0.16	3.9±0.16
		Liver Weight (In Kg)	
Mean±SD	0.026±0.002	0.027±0.002	0.03±0.002
		Gonad Weight (In Kg)	

Parameters	February, 2016	May, 2016	November, 2016
Mean±SD	0.246±0.014	0.271±0.019	0.251±0.008
Condition_Index			
Mean±SD	0.63±0.05	0.65±0.03	0.69±0.03
Gonado-Somatic_Index			
Mean±SD	7.43±0.17	7.9±0.55	6.89±0.4
Hepatosomatic_Index			
Mean±SD	0.75±0.07	0.74±0.04	0.76±0.02
Vitellogenin (ng/ml)			
Mean±SD	398.67±43.28	1586.67±288.33	268±23.93

REFERENCES

1. Agemian H and Chau ASY. (1975). An atomic absorption method for determination of 20 elements in the lake sediments after acid digestion. Anal. Chem. Acta. 80: 61-66.

2. Anu, Upadhyaya SK, Bajpai A. (2011). Heavy Metal Analysis of Various Water Bodies Located in and around Bhopal, M.P.(India). International Journal of Environmental Science and Development. 2(1): 27-29.

3. APHA. (1998). Standard methods for the examination of water and waste water 20th Ed., New York

4. Bajpai A, Vyas A, Verma N and Mishra DD. (2009). Effect of idol immersion on water quality of twin Lakes of Bhopal with special reference to heavy metals. Poll. Res. 28(3): 433-438.

5. Dwivedi VK, Choubey VK. (2008). Proceeding of Taal 2007: The 12th Lake conference. 2110-2125.

6. E. Burzawa-Gerard, A. Dumas-Vidal. (1991). Effects of 17β-estradiol and carp gonadotropin on vitellogenesis in normal and hypophysectomized European silver female eel (Anguilla anguilla L.) employing a homologous radioimmunoassay for vitellogenin, General and Comparative Endocrinology. 84(2): 264-276.

7. Hansen PD, Dizer H, Hock B, Marx A, Sherry J, McMaster M and Blaise C. (1998). Vitellogenin – a biomarker for endocrine disruptors, TrAC Trends in Analytical Chemistry. 17(7): 448-451.

8. Hossain S, Md. Miah I, Md. Islam S and Md. Shahjahan. (2016). Changes in hepatosomatic index and histoarchitecture of liver in common carp exposed to organophosphate insecticide sumithion. Asian J. Med. Biol. Res. 2(2): 164-170

9. Johnson WW and Finley MT. (1980). Handbook of Acute Toxicity of Chemicals to Fish and Aquatic Invertebrates, Resource Publication 137. U.S. Dept. of Interior, Fish and Wildlife Service.Washington, DC.

10. Mosely MP. (1983). Variability of water temperatures in the braided Ashley and Rakia rivers (New Zealand), New Zealand Journal of Marine and Freshwater Research. 17: 331-342.

11. Papoulias DM, Chapman D and Tillitt DE. (2006). Reproductive condition and occurrence of intersex in bighead and silver carp in the Missouri River. Hydrobiologia. 571: 355-360.

12. Phiri O, Mumba P, Moyo BHZ and Kadewa W. (2005). Assessment of the impact of industrial effluents on water quality of receiving rivers in urban areas of Malawi, International Journal of Environmental Science and Technology. 2(3): 237-244.

13. Radke RJ and Kahl U. (2002). Effects of a filter-feeding fish [silver carp, *Hypophthalmichthys molitrix* (Val.)] on phyto- and zooplankton in a mesotrophic reservoir: results from an enclosure experiment. Freshwater Biology. 47(12): 2337-2344.

14. Robison HW and Buchanan TM. (1988). Fishes of Arkansas. University of Arkansas Press, Fayetteville, AK.

15. Singh PK and Shrivastava P. (2016). Assessment of water quality of Upper Lake, Bhopal (M.P). International Journal of Environmental Sciences. 7(2): 164-173.

16. Srivastava N, Harit G and Srivastava R. (2009). A study of physico-chemical characteristics of lakes around Jaipur, India, Journal of Environmental Biology. 30(5): 889-894.

17. Sumpter JP and Jobling S. (1995) Vitellogenesis as biomarker for estrogenic contamination of the aquatic environment. Environ. Health Prep. 103(7): 3-8.

18. Tay CK. (2007). Chemical characteristics of groundwater in the Akatsi and Ketu Districts of the Volta Region, Ghana, West African Journal of Applied Ecology. 11: 3-25.

19. US Geological Survey. (2018). Nonindigenous Aquatic Species Database. Gainesville, Florida. Accessed on 1/23/2018.

20. Wanner GA and Klumb RA. (2009). Length-Weight Relationships for Three Asian Carp Species in the Missouri River. Journal of Freshwater Ecology. 24(3): 489-495.

21. WHO (2018) Children's environmental health, Environmental risk, Endocrine Disrupting Chemicals (EDCs) http://www.who.int/ceh/risks/cehemerging2/en/accessed on 23-01-2018.

Chapter 4

A Study on Anxiety Related Behavioural Alterations in Zebrafish

Whidul Hasan, Manisha, Juli Jain and Deepali Jat*

²*Neuroscience Lab, School of Biological Sciences,*
Dr. Harisingh Gour Central University, Sagar – 470 003, M.P.
*E-mail: deepalipunia@gmail.com

In neurobehavioral research, animal model have played an important role in yielding experimental data as well as in the development of new theories of brain pathogenesis. The aim of study is to evaluate the behavioural alterations in zebrafish induced by glutamic acid, also known as monosodium glutamic acid which widely used as a food additives. Researchers are always trying to develop novel animal models to understand fundamental features of physiological, psychological and behavioural disorders. Zebrafish (Danio rerio) is considered to be a good model to study neurobehavioural changes because fish does not possess the complex behavioural phenotype exhibited by many other animal models and therefore is an excellent alternative to mammalian like animal model. In this study, behaviour of zebrafish was evaluated by glutamate excito-toxicity, altered response in their top/bottom entries and thigmotaxic response (an organism's response to the stimulus of contact or touch). The result of this experiment suggest behavioural changes in zebrafish induced due to exposure of glutamic acid lead to degenerative changes and pathological consequences in the zebrafish brain.

Keywords: Zebrafish, Anxiety, Thigmotaxic, Top/bottom dwelling test and Neurobehavioural study.

Introduction

The behavioral science encloses many regulations and interrelationships between the organisms in the living world (Devereux G, 1967). Behavior is one of the highest levels of research in biology, which is the most complex function of the nervous system. The neuronal system of an organism is affected by adverse psycho-social factors, diseases, stress, drugs and environmental changes which lead to behavioural disturbances(Milgram S, 1963, McEwen B.S. 1993, Chrousos G.P., 2009). Dopamine, acetylcholine, serotonin and glutamic acid are neurotransmitters which play an important role in behaviour regulation. Any imbalance in the life cycle of the neurotransmitter may change the behavioural responses leading to psychiatric illness and many other brain disorders. Many behavioural paradigms of animals have been extensively used as a model system to figure out about physiology and the fundamental features of human behavior (Kumar S. *et al.*, 2015).

Glutamic acid is the most abundant excitatory neurotransmitter in the vertebrate's central nervous system, which involved in many brain functions such as learning, memory and development of nervous system. However various studies suggest that monosodium glutamate (MSG) may exert the toxic effects on the humans and also experimental animals model (Redding *et al.*, 1971 and Yang J.L. *et al.*, 2011). According to National Institute of Mental Health, the excitatory neurotransmitters are encouraging the flow of signals between neuron-cells, which support in the perfect functioning of the cells. However some mental disorders like as depression and schizophrenia, are cause in the humans because of an impotence of the central nervous system to use glutamate (Shekhar A.*et al.*, 2001, Zarate C *et al.*, 2002 and Egan MF *et al.*, 2004). Outside the biomedical scientific community glutamic acid is probably known as "Monosodium glutamate" that is used as a flavour or taste enhancer in food. Over activity of glutamate have been considered to be potent neurotoxic (Thomas R. J., 1995). During this investigation, glutamic exposed fishes were analyzed and behaviuor were recorded in terms of thigmotaxic, top/bottom dwelling behavior to assess the change in behavioral response that lead to pathophysiological alterations in the brain of zebrafish.

Experimental Setup and Methodology

Animals

Adult Zebrafish of heterogeneous genetic make-up of mixed gender were obtained from commercial authorized distributor (Aqua–Forest-Aquarium, Sagar, M.P.) that are experimentally naive and kept 15 days to acclimatize to the laboratory environment before being used for experiment. Laboratory conditions were maintained under a constant photoperiod of 10:14 (Light: Dark) hours dark and light cycle to avoid the changes related to circadian cycle and fed to the chopped prawns at the rate of 0.04 gm/gm body weight given *ad libitum*. De-chlorinated water with a dissolved oxygen concentration of 6.8 ± 0.2 mg/L, total dissolved solute of 138 ± 2, temperature 26 ± 2 °C, pH 7.3 was used and the room temperature was maintained at 26 ± 2 °C. Fishes were kept at a maximum density of 04 individuals in polypropylene recirculation aquaria (L; 30 cm, H; 30 cm, W; 15 cm). The glutamic acid was purchased from HiMedia Laboratories Pvt. Ltd. Mumbai, India.

Drug Administration

Glutamic acid was administered by intraperitoneally to the zebrafish with concentration 0.03mg/gm (b.w.). The dose of drug was calculated on the base of body weight of the zebrafish.

Test Apparatus and Procedure

Behavioural acquisition was performed in a custom made imaging chamber (L; 65cm, H; 62cm, W; 46cm). This imaging chamber made of wooden ply pieces, which are installed in three sided wall to prevent any external stimulus, yellow sheets of paper were placed every side of wall to ensure a uniform background for the video analysis. For top and bottom dwelling test (Novel tank diving test) behaviour measurement apparatus was made of a trapezoidal glass tank 23.9cm x 6.1cm(length x width) in bottom and 28.9cm x 7.4cm(length x width) and 15.1cm height and 15.9cm (height) in diagonal side 12.1cm height were filled with 1.5L water. Open field behaviour measurement apparatus was made of transparent plastic chamber with dimensions 29cm x 37cm x 18cm (width x length x height) containing 10cm of water.

The behavioural test was performed during the same time frame each day (between 10:00am and 4:00pm). During sessions of behavioural measurement, apparatus was filled with same water of home-tanks conditions and all the experiments were performed in a stable surface with minimize environmental distractions.

Treatment and Behavioural Studies

Experimental groups were made for behavioural measurements of fishes. The experimental fishes were divided into two groups; group Ist served as control and group IInd served as glutamate treated group, in each group four fishes (mixed gender) were used in behavioural recordings. Each groups given with chopped prawns *ad libitum* and supplied aeration daily. Zebrafish were transferred from their housing tanks to an experimental tank (made of translucent plastic) an hour before starting the experiment. There was no drug exposure in either the housing chamber. All the control group fishes used only once for behavioural measurement. Swimming behaviour was assessed by the debt video capture software was recorded for 6 min of exploratory behaviour viz top, bottom and open field test. Before each behavioural measurement, water in the behavioural chamber was always replaced with fresh water of same housing tank and maintained the home condition for preventing the behavioural distraction from the environmental stress.

Analysis of Behavioural Test

The behavior analysis was performed by using ImageJ software (v 1.6.0_24 32 bit nih.gov) to track the behavior activity of zebra-fish. This method involved video processing steps and using them for automated tracking of Zebrafish. Video was recorded by using debut video recording software in AVI format at the rate 30 frames/sec, recorded video were segmented AVI files and video was saved as uncompressed AVI file which was significantly bigger in size (Nema S, 2016). The segmented video were then used to analyze all exploration across the six minutes by using ImageJ, such as total time spent in top and bottom in top/bottom test and the thigmotaxic behavior in open field test.

Testing apparatus were placed in behavioral chamber and proper light source was provided. The apparatus was within the camera vision range and it was used to record the locomotion and behavioural activity of the fish. The webcam (Odyssey Co.), were connected to the laptop to store the data and analyze videos using ImageJ automated tracking.

Result

During this study intraperitoneal injection of glutamic acid (0.03mg/gm b.w.) was given to fishes, to observe behavior alteration caused by its excitotoxicity. Glutamic acid is released during some metabolical activities in the brain and also play an important role in glutamatergic transmission. The brain cell dysfunctioning or changing in glutamatergic transmission may be responsible to induce anxiety into the fish. Hence, fish show abnormal behavior during experimentation in novel tank diving test and thigmotaxic test.

Novel Tank Test

In the novel tank test experimentation, glutamic acid has been administered intraperitoneally to the fishes at the rate of 0.03mg/gm b.w., the manually tracked 6 minute data were observed and wonderstruck effects were seen in novel tank diving test. The analysis pattern of these behavioural tests were; top and bottom movements, transition between these vertical areas across time, total entry to the top and total entry to the bottom, erratic movements as well as freezing bouts. We have analyzed the duration in top area and bottom area of treated and control group. Significant reduction was observed in latency to enter the top by the fish in treated group during six minutes data recording and the data were analyzed manually with the help of computer generated 2D traces. These behavioral profiles showed in computer generated 2D traces in heat map analyzed image of zebra fish swimming [Figures 4.1(a) and (b)] 04 fishes in each group have been used for obtaining statistically significant data.

The change in behavior is noted with the help of change in the top and bottom dwelling test. Control group fishes shows their natural preference of swimming as top dwelling behaviour while glutamic acid treated fish shows reduction in the top dwelling frequency and fish moves only in the bottom area of the novel tank. Fishes spent maximum time in the bottom of the tank and spent less time in top of the novel tank suggests the anxiety state of fish due to excite-toxic action of glutamic acid. This change suggests the anxiety is higher in the group, in which glutamic acid is administered intraperitoneally. However, extent of certain amount of glutamic acid may be essential for signal transduction processes in the cell but excessive release can cause deleterious changes which may affect neurobehavioural responses.

Thigmotaxic Response

The fishes are kept under open field experimental chamber to observe thigmotaxic behaviour of fishes. Control group fishes during the study tend to remain at center in the chamber rather than on edges of the chamber. However glutamic acid exposed fishes tend to move near walls of the experimental chamber maximum time.

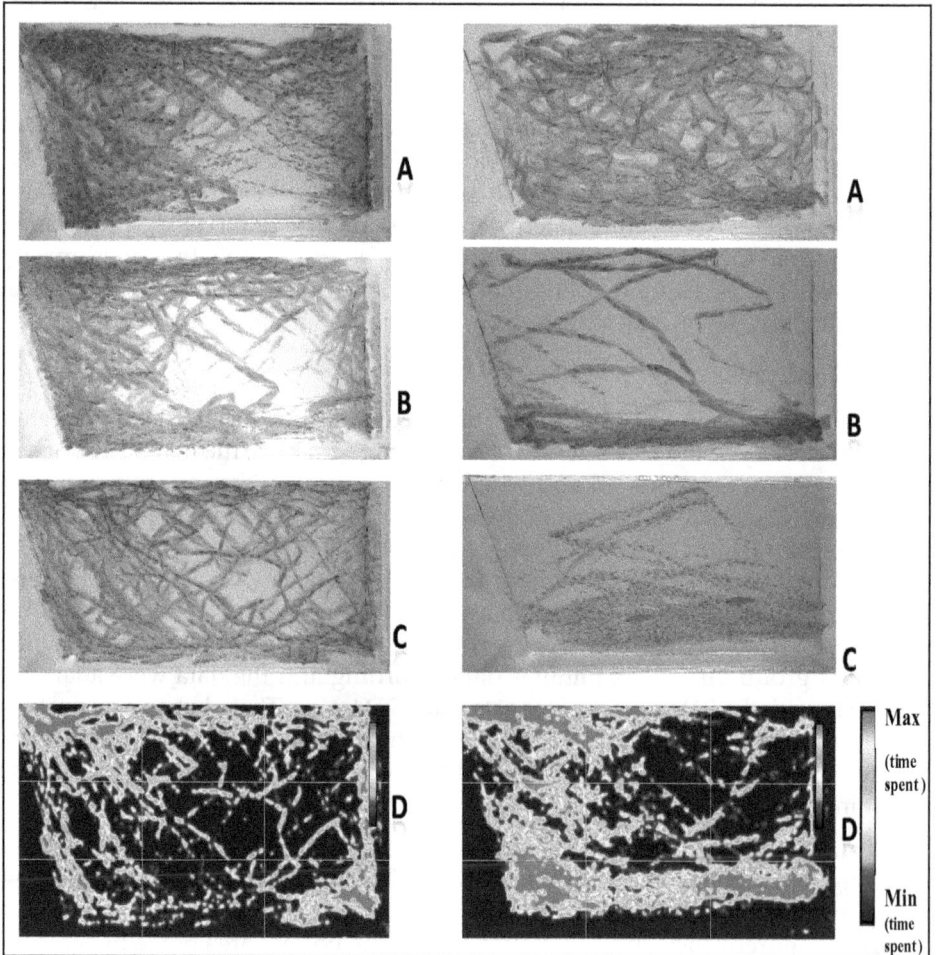

Figure 4.1(a)
Group (I): Control

Figure 4.1(b)
Group (II): Treated

Figure 4.1: The Heat Map.

Figure 4.1(a) shows; the movement of control group, fishes spent maximum time on top of the tank and Figure 4.1(b) shows and the movement of treated group, the glutamic acid treated fishes spent maximum time in the bottom of the tank. The wide red area in treated group tank suggests fish are under stress which indicated the changes in their neurobehavioral response.

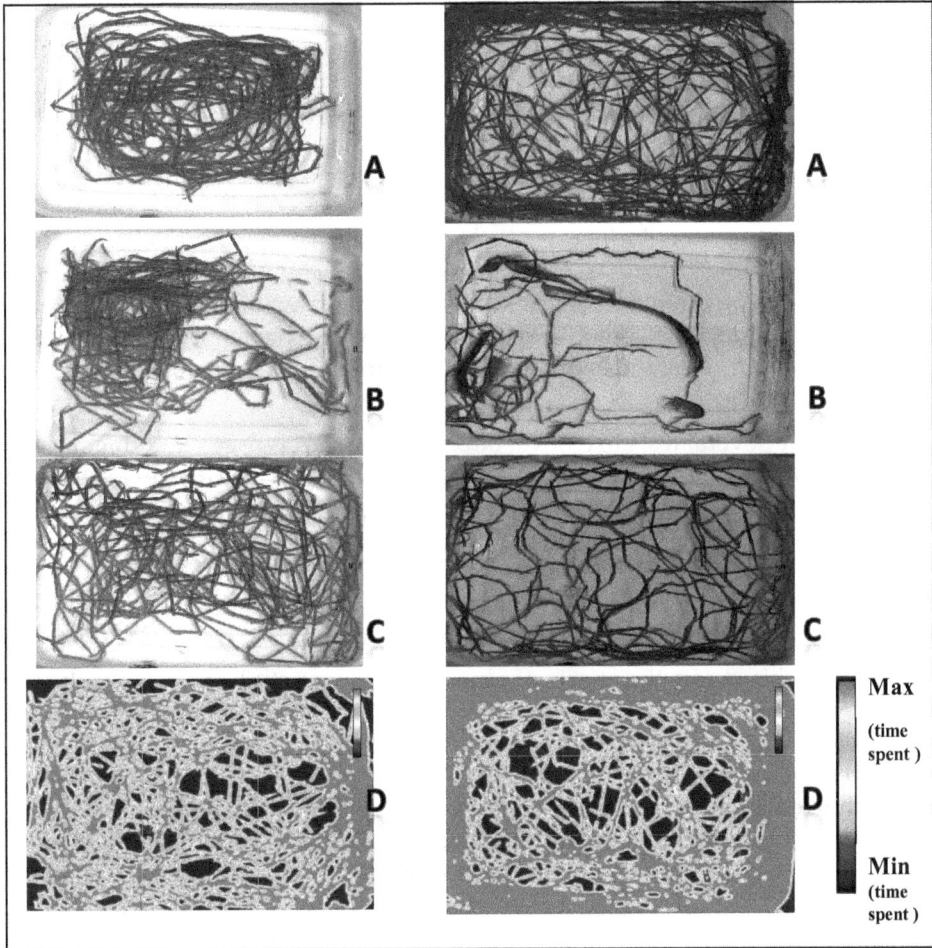

Figure 4.2(a)
Group (I): Control

Figure 4.2(b)
Group (II): Treated

Figure 4.2: The Heat Map.

Figure 4.2(a) shows, the movement of the control group fishes spent maximum time in the center of chamber and heat map Figure 4.2(b) shows, the movement of the glutamic acid treated fishes spent maximum time near the wall of the chamber. The wide red area in treated group, Figure 4.2(b); suggests fishes are under stress and anxiety which indicated that change in their anxiety related neurobehavioral response.

Therefore the mean time spent at nearest wall of the chamber is greater in glutamic acid treated group. The results suggest change in neurobehavioural alteration is due to stress induced with high glutamic acid concentration in fish.

Discussion

Recent advancements in the field of neurobehavioural research provide significant contribution of zebrafish as a tool to study the neurobehavioural changes and its relation with pathophysiological response. Introducing zebrafish in neurobehavioural research may be the advanced approach in understanding behavioural and physiological phenotypes of stress and anxietyKauleff A. V. 2007, Egan R. J et al., 2009 and Stewart A et al., 2011and Kumar S. 2015). Behavioral studies on zebrafish are generally increased exponentially. To the casual observer, the natural behavior of a fish may seem unrelated to that of human. However zebrafish and human share a common ancestry, and as such there are some behaviors that have been conserved over the course of evolution. While we have not yet investigated natural behaviour of zebrafish in much detail compare to other animals (Gerlai R, 2011).

There are several tests like novel tank diving test, scototaxis and thigmotaxic response that are used specially to test anxiety and stress related changes in behavior of zebrafish like novel tank test, scototaxis and thigmotaxic (Sackerman et al., 2010). Those of human if we induce stresses in the zebrafish, altereted fear responses, cause anxiety and stress increased cortisol levels can be observed (Egan et al., 2009 and Champagne D. L., 2010). Thus zebrafish considered as a useful animal model to study of the anxiety, stress and new drug screening.

During this study, the effects of glutamic acid was studied in zebrafish. The daily dose administered to the fish intraperitoneally was 0.03mg/g b.w. behaviuoral parameters like novel tank diving test, thigmotaxic response of the fish were measured during the study with the help of software Image J. Glutamic acid treated fishes showed differences with control group fishes. Each fish subjected to the drug was found to spent maximum time attached to the walls of the tank in thigmotaxic behavior. Also fish found to spent greater time in the bottom of the tank in novel tank diving test in glutamic acid treated group. Our results demonstrated that when a zebrafish is exposed in the stressed environments with glutamic acid it showed anxiety like behavioural changes which we measured with the help of novel tank test and thigmotaxic test (open field test).

Thigmotaxic behavior test is showing by different of the species like sh, rodents and humans (Champagne et al., 201, Kallai J et al., 2007 and Schnorr et al., 2012). The behaviour test show the behavioural endpoints behavior, in which animal are staying closed to the walls of experimental chamber (Heisler L et al., 2007 and Jain A. 2012). Hence, zebrafish can be the good a model to study anxiety like behavior changes (Levin E. D. 2007, Cachat J. 2010, Wong k, 2010 and Stewart A, 2012). In accordance with the above studies evidences suggest that if similar environmental conditions provided to the rats/mice and also to the zebrafish, they cause the similar behaviuoral responses (Champagne D L, 2010, Wong K, 2010 and Stewart a. 2012). P Evidences from the previous studies suggest that fish have some forms of the memories (Riedel G 1998, Chandroo K.P, 2004, Perera D, 2004 and Perera D. 2004) as the rodents (Eilam D. 1998 and Dvorkin A, 2008) and they used their memories to guide to having themselves in the protecting area in the novel environment condition (Rosemberg D B 2011). In the novel tank, zebrafish shown the erratic movements and in the dangerous/fear conditions, the hyper arousal behaviour shown by the rodents, both conditions show similarities in the behaviour (Egan R J 2009)., Thigmotaxic test also known as "wall hugging" test which is one of the mostly used behavioral test in preclinical studies employing rodent models. Those animal models are used in thigmotaxic behavior test, which greatly avoid the center area of the test chamber and move or stay near to the wall of the novel chamber (Schnorr S. 2012 and Sharma S. 2009).

Anxiety related effects of the glutamic acid were observed in the present investigation in the zebrafish. Thigmotaxic test in zebrafish is very much used in the pharmacological studies (Richendrfer H, 2012 and Richendrfer H, 2012). As the glutamic acid treated group zebrafish are swimming in the centre of the open field or thigmotaxic test chamber which indicated an anxiety and stressed behaviour, like as α-fluoro methyl histidine also shows an anxiolytic effects by increasing swimming time in the centre of the chamber[32] but acute ethanol exposure reduces the erratic movements in the thigmotaxic/open field test with zebrafish (Egan R j, 2009). The results of the (Kumar, Joshi *et al.*, 2015) suggest, zebrafish is a validated animal model for the study of anxiety and stress. The zebrafish has many quantifiable behavioural, neuro-pathological and physiological responses similar to humans[37].

Zebrafish has a similar basic organization of the brain components to the humans; make it useful in the study of neurobehavioural response (Mortez J A, 2007, Tomasiewicz H G (2002 and Panula P, 2010). Many structural systems of human brain are showing similarities with zebrafish brain systems like aminergic system (Sallinen V, 2009 and Sager J J, Bai Q, Burton E A, 2010) and cortisol is a primary stress hormone in the both human and zebrafish (Alsop D M,Vijayan M, 2009). Therefore altered behavior of zebrafish due to little increase in the concentration of neurotransmitter glutamate can be a disturbance in the psysiological responses of brain cells.

Conclusion

In this study we addressed whether there were abnormalities in vulnerable brain by examining top/bottom dwelling and thigmotaxic response in glutamic acid induced zebrafish. These observations may be important subjects for future research on zebrafish behaviour. Indeed, the understanding of glutamic acid induced behaviour might open interesting perspectives in terms of understanding behavioural neuroscience. Recent findings suggest significant contribution of *Danio rerio* in the studies related to neuropathological changes. Thus present study may provide significant approach towards developing scientific view of recognizing the zebra fish as a model to anxiety related behavior study and also suggest the deleterious changes in anxiety behaviour of the zebrafish with glutamic acid.

Acknowledgements

The author is thankful to the facilities provided by Madhya Pradesh Council of Science and Technology, Bhopal, M.P.

Author Contributions

Idea and conceptualization by DJ; Experimental validation by WH and DJ; Formal data analysis by WH; Writing – Original Draft by DJ and MN; Writing – Review and Editing by MN and JJ; Data visualization, supervision, project administration and funding acquisition by DJ.

Here, DJ: Deepali Jat; WH: Whidul Hasan; MN: Manisha Nahar; JJ: Juli Jain.

REFERENCES

1. Devereux G :From anxiety to method in the behavioral sciences, Walter de Gruyter GmbH and Co KG. 1967.

2. Milgram S :Behavioral Study of obedience. The Journal of abnormal and social psychology. 1963 ; **67**(4): 371.

3. McEwen B S, Stellar E :Stress and the individual: mechanisms leading to disease. Archives of internal medicine. 1993; **153**(18): 2093-2101.

4. Chrousos G P : Stress and disorders of the stress system. Nature Reviews Endocrinology. 2009;**5**(7): 374-381.

5. Kumar S, Joshi H, Bahuguna P, Kumar R :Research Article A study on behaviour of zebra fish. 2015.

6. Redding T W, Schally A V, Arimura A, Wakabayashi I: Effect of monosodium glutamate on some endocrine functions.Neuroendocrinology.1971; **8**(3-4): 245-255.

7. Yang J L, Sykora P, Wilson D M, Mattson M P, Bohr V A: The excitatory neurotransmitter glutamate stimulates DNA repair to increase neuronal resiliency. Mechanisms of ageing and development. 2011;**132**(8): 405-411.

8. Shekhar A, McCann U, Meaney M, Blanchard D, Davis M, Frey K, Liberzon I, Overall K, Shear M, Tecott L: Summary of a National Institute of Mental Health workshop: developing animal models of anxiety disorders. Psychopharmacology-Berlin.2001;**157**(4): 327-339.

9. Zarate C, Quiroz J, Payne J, Manji H: Modulators of the glutamatergic system: implications for the development of improved therapeutics in mood disorders. Psychopharmacology bulletin. 2002;**36**(4): 35-83.

10. Egan M F, Straub R E, Goldberg T E, Yakub I, Callicott J H, Hariri A R, Mattay V S, Bertolino A, Hyde T E, Shannon-Weickert C: Variation in GRM3 affects cognition, prefrontal glutamate, and risk for schizophrenia. Proceedings of the National Academy of Sciences of the United States of America. 2004; **101**(34): 12604-12609.

11. Thomas R J : Exitatory Amino Acids in Health and Disease. Journal of the American Geriatrics Society. 1995;**43**(11): 1279-1289.

12. Nema S, Hasan W, Bhargava A, Bhargava Y :A novel method for automated tracking and quantification of adult zebrafish behaviour during anxiety. Journal of neuroscience methods. 2016; **271**: 65-75.

13. Kalueff, A V: Neurobiology of memory and anxiety: from genes to behavior. Neural plasticity **2007**.

14. Egan R J, Bergner C L, Hart P C, Cachat J M, Canavello P R, Elegante M F, Elkhayat SL, Bartels B K, Tien A, Tien D H: Understanding behavioral and physiological phenotypes of stress and anxiety in zebrafish.Behavioural brain research. 2009; **205**(1): 38-44.

15. Stewart A, Wu N, Cachat J, Hart P, Gaikwad S, Wong K, Utterback E, Gilder T, Kyzar E, Newman A : Pharmacological modulation of anxiety-like phenotypes in adult zebrafish behavioral models. Progress in Neuro-Psychopharmacology and Biological Psychiatry. 2011;**35**(6): 1421-1431.

16. Gerlai R :Using zebrafish to unravel the genetics of complex brain disorders. Behavioral Neurogenetics, Springer. **2011;** 3-24.

17. Sackerman J, Donegan J J, Cunningham C S, Nguyen N N, Law K, Long A, Benno R H, Gould G G: Zebrafish behavior in novel environments: effects of acute exposure to anxiolytic compounds and choice of Danio rerio line. International journal of comparative psychology/ISCP; sponsored by the International Society for Comparative Psychology and the University of Calabria. 2010; **23**(1): 43.

18. Champagne D L, Hoefnagels C C, de Kloet R E, Richardson M K :Translating rodent behavioral repertoire to zebrafish (Danio rerio): relevance for stress research. 2010.

19. Kallai J, Makany T, Csatho A, Karadi K, Horvath D, Kovacs-Labadi B, Jarai R, Nadel L, Jacobs J W :Cognitive and affective aspects of thigmotaxis strategy in humans. Behavioral neuroscience. 2007;**121**(1): 21.

20. Schnörr S, Steenbergen P, Richardson M, Champagne D: Measuring thigmotaxis in larval zebrafish.Behavioural brain research.2012;**228**(2): 367-374.

21. Heisler L, Zhou L, Bajwa P, Hsu J, Tecott L :Serotonin 5-HT2C receptors regulate anxiety-like behavior. Genes, Brain and Behavior. 2007; **6**(5): 491-496.

22. Jain A, Dvorkin A, Fonio E, Golani I, Gross C T: Validation of the dimensionality emergence assay for the measurement of innate anxiety in laboratory mice. European Neuropsychopharmacology. 2012; **22**(2): 153-163.

23. Levin E D, Bencan Z, Cerutti D T :Anxiolytic effects of nicotine in zebrafish. Physiology and behavior. 2007; **90**(1): 54-58.

24. Cachat J, Stewart A, Grossman L, Gaikwad S, Kadri Chung K M, Wu N, Wong K, Roy S:Measuring behavioral and endocrine responses to novelty stress in adult zebrafish. 2010 ;**5**(11): 1786-1799.

25. Wong K, Elegante M, Bartels B, Elkhayat S, Tien D, Roy S, Goodspeed J, Suciu C, Tan J, Grimes C :Analyzing habituation responses to novelty in zebrafish (Danio rerio). Behavioural brain research. 2010;**208**(2): 450-457.

26. Stewart A, Gaikwad S, Kyzar E, Green J, Roth A, Kalueff A V: Modeling anxiety using adult zebrafish: a conceptual review. Neuropharmacology. 2012;**62**(1): 135-143.

27. Riedel G :Long-term habituation to spatial novelty in blind cave fish (Astyanax hubbsi): role of the telencephalon and its subregions. Learning and Memory. 1998;**4**(6): 451-461.

28. Chandroo K P, Duncan I J, Moccia R D :Can fish suffer?: perspectives on sentience, pain, fear and stress. Applied Animal Behaviour Science. 2004; **86**(3): 225-250.

29. Perera D, Burt T: Fish can encode order in their spatial map. Proceedings of the Royal Society of London B: Biological Sciences. 2004; **271**(1553): 2131-2134.

30. Perera D, Burt T : Spatial parameters encoded in the spatial map of the blind Mexican cave fish, Astyanax fasciatus. Animal Behaviour. 2004; **68**(2): 291-295.

31. Eilam, D, Golani I :Home base behavior of rats (Rattus norvegicus) exploring a novel environment. Behavioural brain research. 1989; **34**(3): 199-211.

32. Dvorkin, A, Benjamini Y, Golani I: Mouse cognition-related behavior in the open-field: emergence of places of attraction. PLoS computational biology.2008;**4**(2): e1000027.

33. Rosemberg D B, Rico E P, Mussulini B H M, Piato Â L, Calcagnotto M E, Bonan C D, Dias R D, Blaser R E, Souza D O, Oliveira D L de :Differences in spatio-temporal behavior of zebrafish in the open tank paradigm after a short-period confinement into dark and bright environments PloS one.2011; **6**(5): e19397.

34. Sharma S, Coombs S, Patton P, de Perera T B: The function of wall-following behaviors in the Mexican blind cavefish and a sighted relative, the Mexican tetra (Astyanax).Journal of Comparative Physiology A.2009;**195**(3): 225-240.

35. Richendrfer H, Pelkowski S D, Colwill R M, Créton R: Developmental sub-chronic exposure to chlorpyrifos reduces anxiety-related behavior in zebrafish larvae. Neurotoxicology and teratology.2012; **34**(4): 458-465.

36. Richendrfer H, Pelkowski S D, Colwill R M, Creton R: On the edge: pharmacological evidence for anxiety-related behavior in zebrafish larvae. Behavioural brain research. 2012 ; **228**(1): 99-106.

37. Moretz J A, Martins E P, Robison B D: Behavioral syndromes and the evolution of correlated behavior in zebrafish. Behavioral ecology. 2007; **18**(3): 556-562.

38. Tomasiewicz H G, Flaherty D B, Soria J, Wood J G: Transgenic zebrafish model of neurodegeneration. Journal of neuroscience research.2002;**70**(6): 734-745.

39. Panula P, Chen Y C, Priyadarshini M, Kudo H, Semenova S, Sundvik M, Sallinen V :The comparative neuroanatomy and neurochemistry of zebrafish

CNS systems of relevance to human neuropsychiatric diseases. Neurobiology of disease. 2010; **40**(1): 46-57.

40. Sallinen V, Torkko V, Sundvik M, Reenilä I, Khrustalyov D, Kaslin J, Panula P: MPTP and MPP+ target specific aminergic cell populations in larval zebrafish. Journal of neurochemistry.2009;**108**(3): 719-731.

41. Sager J J, Bai Q, Burton E A: Transgenic zebrafish models of neurodegenerative diseases. Brain Structure and Function.2010;**214**(2-3): 285-302.

42. Alsop D M,Vijayan M : Molecular programming of the corticosteroid stress axis during zebrafish development Comparative Biochemistry and Physiology Part A: Molecular and Integrative Physiology. 2009 **153**(1): 49-54.

Chapter 5

Screening for Production of Bioactive Molecules from Actinomycetes in some Areas of Shahdol District

Vinita Singh and Bharat Sharan Singh

Department of Biotechnology
Pt. S.N. Shukla Govt. P.G.College,
Shahdol, Madhya Pradesh
E-mail: vinita11chandel@gmail.com

Total four different soil samples were collected from different places of Shahdol. These soil samples are further used for isolation of Actinomycetes from which screening of antimicrobial activity of Actinomycetes was performed. A total number of fifteen Actinomycetes were isolated from different soil samples of Shahdol region. The isolated Actinomycete strain were then screened and selected strains are identified. The cultural and physiological characteristics of the strain identified the strain as a member of the genus Streptomyces. The maximum number of Actinomycetes was isolated from Pandav nagar and College Garden soil samples, where as isolates from soil sample Banganga and Purani Basti had only one isolates of each. This concludes that soil samples College Garden and Pandav Nagar have high population of actinomycetal species and it contains high amount of source of nutrition of Actinomycetes. This may be the reason that these soils contain maximum population of Actinomycetes. Most of the isolated colonies are exhibited pigmentation of light pinkish color or light yellow colors, i.e., these colonies are from genus Streptomyces. Streptomyces are the major source of antibiotics and they are commercially important Actinomycetes. Each Actinomycete imparts different pigmentation on different agar medium. These Actinomycetes gives solely smell of Geosmin which is a volatile compound.

Keywords: Actinomycetes, Streptomyces, Screening.

Introduction

Actinomycete, are a group of bacteria having fungus like morphology. These are a group of unicellular organism, which reproduce either by fission or by means of special spores or conidia. *Actinomycetes* are aerobic, gram positive bacteria having high CG content that form branching haphae and asexual spore which are widely distributed in terrestrial environment (Waskman S.A. 1959). The majority of *actinomycetes* are free living saprophytic bacteria found widely distributed in soil, water and colonizing plant.

Many species of *actinomycetes* occur in soil and are harmless to animals and higher plants, while some are important pathogens, and many others are beneficial sources of antibiotics such as streptomycin.

Actinomycetes contain a cell wall. Flagella grow in and on the substrate. Another internal structure is thallus. One more internal structure is mycelium. *Actinomycetes* can be identified by their branching growth pattern that results in large threadlike structures. These filaments may break apart to form rods or spheroidal shapes, called bacillus. Some *Actinomycetes* can form spores. The *Actinomyces* colony is made up often of two types of mycelium, consisting primarily of vegetative or substrate growth and of secondary aerial or sporogenous growth

When microbial growth is in exponential phase then many intermediate metabolic products are produced. These products are classified into two:

1. Primary metabolites
2. Secondary metabolites

Materials and Methods

Collection of Soil Samples

The soil of shahdol region is mostly black and is very fertile. The main soil types found are medium black, shallow black, mixed red and skeletal. Soil samples were collected from different locations of shahdol district.

Bennett's Agar Medium

Bennets Agar is used for the sporulation and cultivation of Streptomyces species. The medium contains nitrogenous nutrients such as yeast extract, beef extract and casein enzymic hydrolysate. They also serve as sources of carbon and essential growth factors. Dextrose is an energy source.

Potato Dextrose Agar Media

Potato dextrose agar are common microbial growth media made from potato infusion and dextrose. It is the most widely used medium for growing fungi and bacteria

Nutrient Agar Media

Nutrient agar is a general purpose medium supporting growth of a wide range of organisms.

Results and Discussions

Isolation of *Actinomycetes*

Actinomycetes were isolated from soil samples of different locations o Shahdol. The isolates obtained from soil samples are given below:

Sl.No	Location of the Soil Samples	Code	No. of Isolates
1.	College garden	**CG**	6
2.	Pandav Nagar	**PN**	7
3.	Purani Basti	**PB**	1
4.	Bangaga	**BG**	1

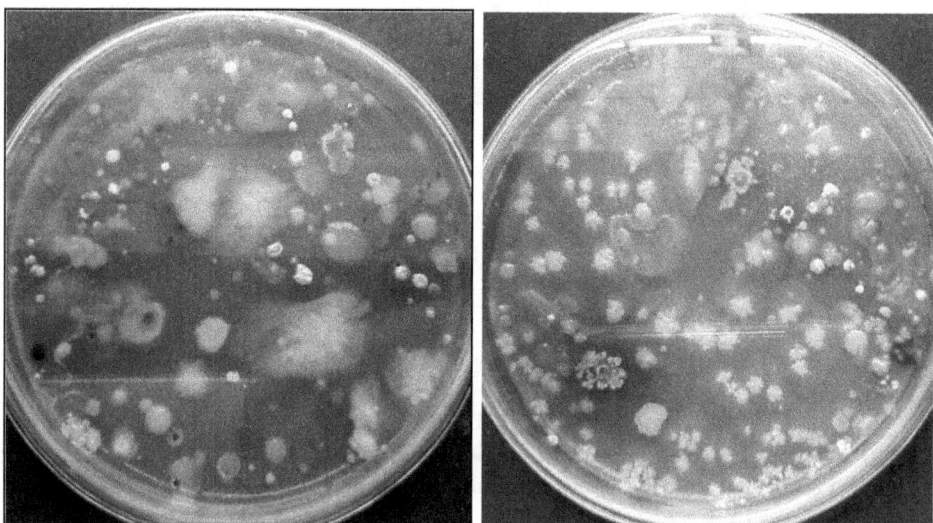

Figure 5.1: Mixed Colonies Showing Actinomycetal Isolates Observed on Plates.

Colonial Characteristics

Sl.No.	Isolated Colony	Colonial Characteristics							
		Size	Shape	Edge	Eleva-tion	Texture	Axial Spore Mass	Colony Reverse	Pigmen-taion
1.	**CG1**	Medium	Round	Irregular	Flat	Chalky	Light yellow	Light yellow	No
2.	**CG2**	Medium	Round	Rhizoid	Raised	Rough	White	White	No
3.	**CG3**	Small	Round	Irregular	Flat	Rough	Pinkish white	Light yellow	Yes
4.	**CG4**	Small	Round	Regular	Raised	Rough	Pinkish white	Cherry red	Yes
5.	**CG5**	Small	Round	Regular	Raised	Rough	White	Cream White	Yes

Sl.No.	Isolated Colony	Colonial Characteristics							
		Size	Shape	Edge	Elevation	Texture	Axial Spore Mass	Colony Reverse	Pigmentaion
6.	CG6	Medium	Round	Irregular	Raised	Velvety	Yellow white	Yellow white	No
7.	PN1	Medium	Round	Regular	Raised	Velvety	Cream white	Yellow	Yes
8.	PN2	Large	Round	Rhizoid	Raised	Rough	White	Yellow	Yes
9.	PN3	Small	Round	Regular	Flat	Chalky	Pinkish white	Cherry red	Yes
10.	PN4	Medium	Round	Regular	Raised	Velvety	Cream White	Light orange	Yes

Antifungal Assay

To determine the antifungal activity of actinomyceties we use mainly five methods

1. Well Diffusion Method

Out of 11 actinomycetal isolates, only two of them showed antifungal activity against two test organisms

Sl.No.	Name of Actinomycetal Isolates	Name of Test Fungi and Zone Diameter									
		H3	Tricho-derma	Fusa-rium	TP	SP	SP2	RH	TP1	Gr	A.s
1.	PN2	-	-	-	-	-	-	-	-	-	-
2.	PN3	-	-	-	-	-	-	-	-	-	-
3.	PN4	-	-	-	-	-	-	-	-	-	-
4.	PN9	-	-	-	-	-	-	-	-	-	-
5.	PN10	-	-	-	-	-	-	-	-	-	-
6.	BG1	16 mm	-	-	-	18 mm	-	-	-	-	-
7.	CG3	-	-	-	-	-	-	-	-	-	-
8.	CG4	-	-	-	-	-	-	-	-	-	-
9.	PB2	-	-	-	-	-	-	-	-	-	-
10.	CG1	15 mm	-	-	-	-	-	-	-	-	-
11.	CG5	-	-	-	-	-	-	-	-	-	-

2. Agar Block Method

In this method 6 actinomycetal cultures were taken against 9 test fungal cultures. But this method was unsuccessful to detect the antifungal activity of actinomycetes cultures. The results are given below:

Sl.No.	Name of the Actinomycetal Cultures	Name of the Test Fungal Cultures								
		H3	Trichodema	H5	SP	SP2	Gr	Aspergillus	TP	Fusarium
1.	PN3	-	-	-	-	-	-	-	-	-
2.	PN9	-	-	-	-	-	-	-	-	-
3.	PN10	-	-	-	-	-	-	-	-	-
4.	CG3	-	-	-	-	-	-	-	-	-
5.	CG2	-	-	-	-	-	-	-	-	-
6.	PB2	-	-	-	-	-	-	-	-	-

3. Cross Section Method

Total 15 Actinomycetal cultures were used against 13 test fungi. Out of which 9 isolates were exhibit antifungal activity against test fungi. The results are given below:

Sl.No.	Name of the Actinomycetal isolates	Name of the Test fungi												
		H3	H5	RH	Asp.	As	SP1	Trichoderma	Fusarium	TP	TP1	Gr	SP	SP2
1.	PN1	-	-	-	-	-	-	-	-	-	-	-	-	-
2.	PN2	-	-	-	-	-	-	-	-	-	-	-	-	-
3.	PN3	-	-	-	-	-	-	-	-	-	-	-	-	-
4.	PN4	+	-	-	-	-	-	+	-	-	+	-	+	-
5.	PN8	+	-	-	-	-	+	-	-	+	-	-	-	-
6.	PN9	-	+	-	-	-	-	+	-	+	+	-	+	
7.	PN10	-	+	+	-	-	+	-	-	+	+	-	+	+
8.	CG1	-	-	-	-	-	-	-	-	-	-	-	-	-
9.	CG2	-	-	-	-	-	-	-	-	-	-	-	-	-
10.	CG3	+	-	-	-	-	-	+	-	-	+	+	-	+
11.	CG4	+	+	-	-	-	+	-	-	-	-	-	-	-
12.	CG5	-	+	-	-	-	-	-	-	-	-	-	-	-
13.	CG6	-	-	-	-	-	-	-	-	-	-	-	-	-
14.	PB2	+	+	-	-	-	+	-	-	+	-	-	-	-
15.	BG1	-	-	-	-	-	+	+	-	+	-	-	+	-

4. Point Inoculation Method

Isolates PN4,PN8, PN9, and CG4 were showed maximum antifungal activity against test fungi SP, because its diameter remain smaller as compare to other test fungi.

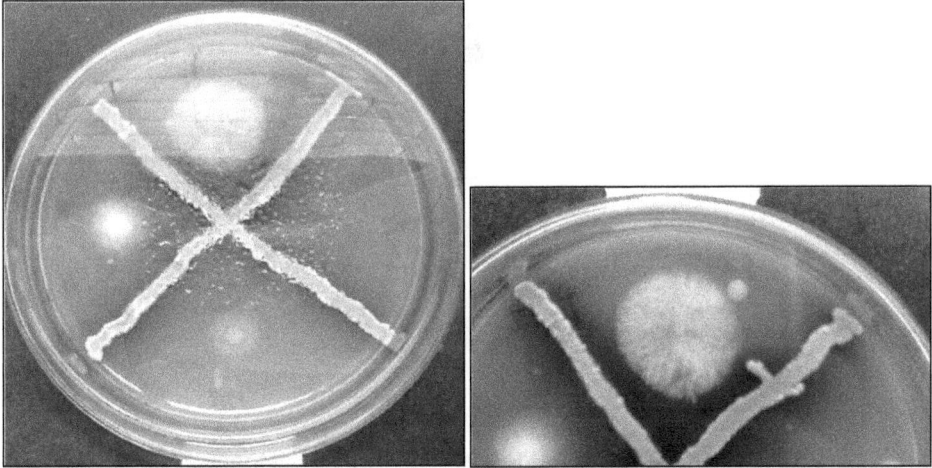

Figure 5.2: Inhibitory Effect Showed by *Actinomycetes* by Secreting Antifungal Metabolite against Test Organism.

Sl.No.	Name of the Actinomycetal Isolates	Fungal growth Diameter (in cm)				
		SP	TP1	Fusarium	Trichoderma	H3
1.	PN2	52 mm	76 mm	75 mm	54 mm	60 mm
2.	PN4	12 mm	30 mm	65 mm	18 mm	38 mm
3.	PN8	18 mm	14 mm	77mm	16 mm	35 mm
4.	PN9	18 mm	19 mm	70 mm	39 mm	62 mm
5.	CG3	20 mm	16 mm	55 mm	25 mm	38 mm
6.	CG4	16 mm	20 mm	74 mm	30 mm	24 mm
7.	PB2	25 mm	20 mm	78mm	76 mm	38 mm
8.	BG1	25 mm	12mm	37 mm	78 mm	15 mm

5. Spot Inoculation Method

This method was also unsuccessful to check the antifungal activity against tests fungi. *Actinomycetes* were not grown properly on SDA agar media due to acidic pH. Antifungal activity was not observed in the absence of secondary metabolites.

Sl.No.	Name of the Actinomycetal Isolates	Name of the Test Fungus			
		Trichoderma	H5	TP	RH
1.	CG6	-	-	-	-
2.	CG1	-	-	-	-
3.	BG1	-	-	-	-
4.	PN3	-	-	-	-

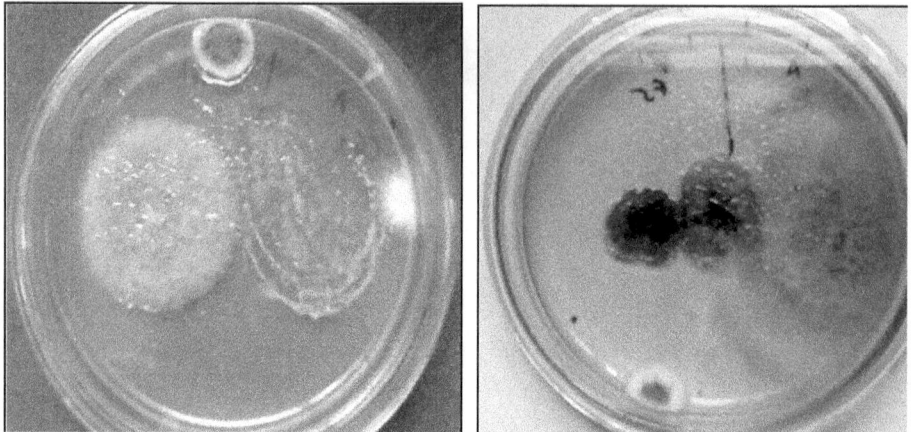

Figure 5.3: Determination of Antifungal Activity by Spot Inoculation Method.

Determining the Antibacterial Activity of Actinomycetes

To determine the antibacterial activity of actinomycetes we use well diffusion method.

Antibacterial Assay

Out of 15 isolates 8 were showed antibacterial activity of *actinomycetes* against test bacteria. All the 8 isolates were exhibit inhibitory effect against gram negative bacteria (*Salmonella* and *Proteus*) whereas 8 isolates were showed zones of inhibition against gram positive bacteria (*Bacillus* and *staphylococcus*). There was no antibacterial activity showed against *Klebshiella* and *Pseudomonas*.

Sl.No.	Name of the Actinomycetal Isolates	Name of Test Bacteria					
		Salm-onella	Klebs-iella	Proteus	Bacillus	Staphylo-coccus	Pseudo-monas
1.	PN1	-	-	17 mm	21 mm	14 mm	-
2.	PN2	-	-	-	-	-	-
3.	PN3	-	-	19 mm	12 mm	-	-
4.	PN4	20 mm	-	28 mm	13 mm	12 mm	-
5.	PN8	-	-	20 mm	22 mm	19 mm	-
6.	PN9	22mm	-	23 mm	-	13 mm	
7.	PN10	-	-	13 mm	-	12 mm	-
8.	CG1	-	-	16 mm	17 mm	16 mm	-
9.	CG2	-	-	-	-	-	-
10.	CG3	-	-	-	-	-	-
11.	CG4	-	-	-	-	-	-
12.	CG5	-	-	-	-	-	-
13.	CG6	-	-	-	-	-	-
14.	PB2	24mm	-	22mm	-	16mm	-
15.	BG1	-	-	-	-	-	-

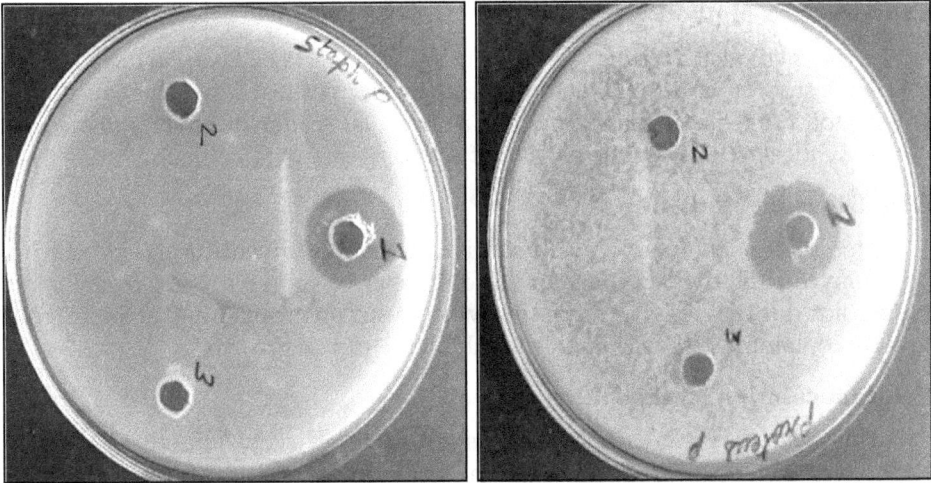

Figure 5.4: Zone of Inhibition are Observed in Well Diffusion Pour Plate Technique.

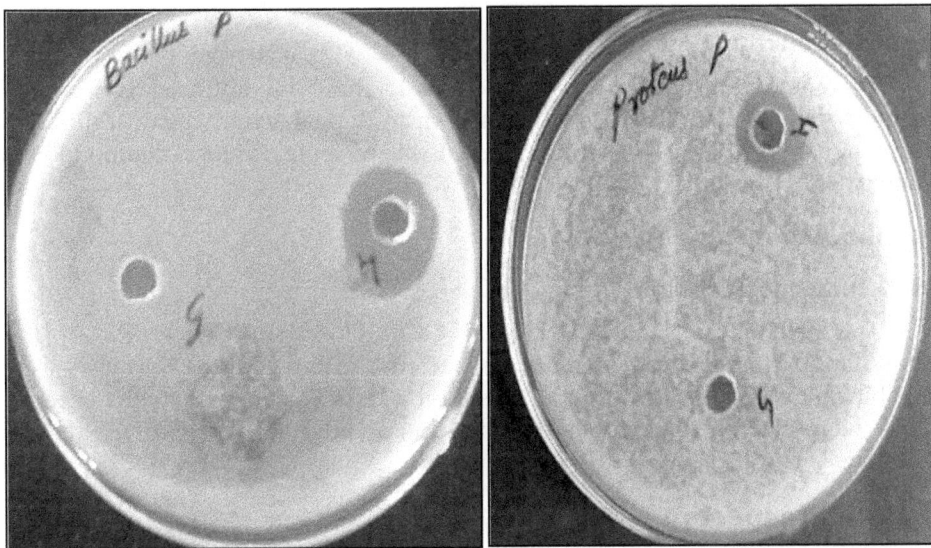

Figure 5.5: Zone of Inhibition were Observed in Pour Plate Well Diffusion Technique.

Conclusions

Antifungal Assay

Well Diffusion Pour Plate Technique

BG1 and CG1 are antifungal compound producing Actinomycetes species. It is concluded on the basis of appearance of zone of inhibition around wells containing supernatant.

Cross-section Method

In this method. the results were concluded that PN4, PN8, PN9, PN10, CG3,CG4, PB2, BG1 showed antifungal activity against different test fungi. the conclusion were made due to having less growth in a particular section or fungal growth may acquiring the shape of outline growth of actinomycetes.

Disc Diffusion Method

This method was given an unsatisfied results for antifungal producing Actinomycetal isolates. Because of large number of spores present in the spore suspension due to this dense fungal growth appears on the plate and actinomycetal growth containing disc was unable to inhibit the fungal growth.

Point Inoculation Method

This method gave satisfactory results at some level. it concludes that PN4, PN8, PN9, CG3, CG4, PB2, BG1 Shows antifungal activity against Test fungi. The isolate which have fungal growth with small diameter on plate that is the best antifungal producer.

Spot Inoculation Method

This method also all most become fails to show any antifungal activity of actinomycetal isolates. Isolates which shows antifungal activity against different test fungi are further used as fungicides. Inspite of using chemical compounds as fungicides.

Antibacterial Activity

Well Diffusion Technique

This method results that PN1, PN3, PN4, PN8, PN9, PN10, CG1 andPB2 isolates shows antibacterial activity against test bacterial culture. All the 8 isolates were exhibit inhibitory effect against gram positive bacteria, Bacillus and Staphylococcus. They were also showed zones of inhibition against gram negative bacteria. Salmonella and Proteus.

Future Prospects

According to these results it was concluded that Actinomycetes have Antimicrobial Activity which can be further used in making new antibiotics against multi drug resistant bacteria and fungi. These antibiotics will have high economic value and will be of great use to human kind. These can be as much cheaper than high expensive chemically antibiotic drugs.

REFERENCES

1   A.Kavitha, M. Vijaylakshmi, P. Sudhakar and G. Narsimha, 2009. Screening of Actinomycete strains for the production of antifungal metabolite. African journal of Microbiology Research vol. 4 (1), pp. 027 – 032.

2   **Denitsa Nedialkova and M. Naidennova, 2004 – 2005.** Screening the antimicrobial activity of actinomycetes strains isolated from antartica. Journal of culture collections volume 4, pp, 29 -35.

3   **Harpreet Sharma and L. Parihar, 2010.** Antifungal activity of extracts obtained from actinomycetes. Journal of yeast and fungal Research vol, 1(10), pp, 197 – 200

4   **Prapassaron Rugthaworn, U. Dilokkunanant, S. Sangchote, N. Piadang and V. Kitpreechavanich, 2007.** A search and improvement of actinomycetes strains for biological control of plant pathogens. Kasetsart J. (Nat. Sci.) 41: 248 – 254.

5   **Raja and P. Prabakarana, 2011.** Actinomycetes and Drug-An Overview. American Journal of Drug Discovery and Development, 1: 75-84.

Chapter 6

Impact of Global Warming on Ecotourism and Environment

Sunita Singh

Department of Zoology
Govt. Girls PG Excellence College, Sagar, M.P.

Ecotourism today is the most vibrant tertiary activity and a multi-billion industry in india. Traditionally known largely for its historical and cultural dimensions, tourism today is highlighted for its immense business opportunities. With its lucrative linkages with transport, hotel industry etc., potential and performance of India's tourism industry needs to be gauged in terms of its socio-economic magnitudes. Global warming is defined as an increase in the average temperature of the Earth's atmosphere, especially a sustained increase great enough to cause changes in the global climate' This paper provides a review of some of the literature which focuses on environmental impacts of ecotourism.

Keywords: *Ecotourism, Global warming, Biodiversity.*

The quality of environment both natural and manmade is essential to ecotourism and is relationship with the environment is complex. It involves many activities that can have adverse environmental effects. Many of these impacts are linked with the construction of general infrastructure such as roads, colonies and tourism facilities, including resorts, hotels *etc.*

The negative impacts of ecotourism development can gradually destroy the environmental resources on which it depends. On the other hand, tourism has a potention to create beneficial effects on the environmental by contributing to environmental protection and conservation. It is a way to taise environmental values and increase the economic importance.

The absorbed solar energy heats our planet, As the rocks, the air, and the seas warm, they radiate "heat" energy (thermal infrared radiatimn). From the surface, this energy travels into the atmosphere where much of it is absorbed by water vapor and long- lived greenhouse gases such as carbon dioxide and methane. 'Global warming is defined as an increase in the average temperature of the Earth's atmosphere, especially a sustained increase great enough to cause changes in the global climate'.

Environmental Impacts of Tourism at Global Level

Loss of Biodiversity

Nature tourism is closely linked to biodiversity and the attractions created by a rich and varied environment. It also can cause loss of biodiversity when land resources are a trained by excessive use and environment and water resources exceed their carrying capacity. This loss of biodiversity means loss of tourism potential.

Depletion of Ozone Layer

Ozone layer which is situated in the upper atmosphere or stratosphere at an altitude of 12-50 kms protects life on earth by absorbing the harmful wavelengths of the sun's UV rays ehich in high doses are fatal to life forms. Refrigerators, air conditioners and propellants in aerosol a pray cans amongst others contain ozone depleting substances.

Improved Environmental Management and Planning

Sound environmental management of as tourism facilities hotels can increase benease benefits to natural areas. But this requires careful planning for controlled development based on analysis of environmental resources of the area. Cleaner production tools can be important for planning and operating facilities in a way minimizes ther environmental impact (UNEP 1995, WTO, 1995).

Ecotourism is referred to as sustainable nature based tourism. It associates tourism in synchronization which nature and offers opportunities for tourists to experience and explore the nature. It also highlights the absolute need of protection of biodiversity and local culture. Ecotourism focuses on visiting areas featuring fragile, pristine, and relatively undisturbed are prime attractions.

Madhya Pradesh is gifted with undisturbed landscapes, forests and wildlife and cultural diversity the state has the largest forest area in the country of which 10000 sq.km is underneath protected areas like national parks and sanctuaries. Since most of its forest area is covered with forest ecotourism have the potential of play an important role in preceding:

1. Unlimited income/higher retained income-even though tourism is poorly developed in Madhya Pradesh, but it can contribute a lot to the economy.

2. Promoting local cuisine and culture-as tourism increases local handicrafts, specialty food items, and other souvenirs can be marketed. Village to village trekking tours can be structured to give a gist of local culture.

3. Employment to local people-ecotourism can provide employment opportunities and job training to segments of society who have been long term unemployed, are disabled and gender- discriminated women (Baked and Bhagvalula 2010),

It is relatively new segment in India. It involves natural areas without disturbing the fragile ecosystem. Eco tourism generates wealth for local people, who in turn take measures to conserve and protect the environment and natural resources. India with its natural diversity is one of the pristine places in the world for eco tourism.

The Himalayan region, Kerala, Northeast, Andaman and Nicobar Islands and Lakshadweep islands the western and Eastern Ghats are some of the hot spots for eco tourism in India. India has some of the best wildlife reserves in the World, rich in flora and fauna. Ecotourism is more than a catch phrase for nature loving travel and recreation.

Eco-tourism is consecrated for preserving and sustaining the diversity of the world's natural and cultural environments. It accommodates and entertains visitors in a way that is minimally intrusive or destructive to the environment and sustains and supports the native cultures in the locations it is operating in. Responsibility of both travelers and service providers is the genuine meaning for eco-tourism. Eco-tourism also endeavors to encourage and support the diversity of local economies for which the tourism telaed income is important. With support from tourists, local services and producers can compete with larger, foreign companies and local families can support themselves. Besides all these, the revenue produced from tourism helps and encourages governments to fund conservation projects and training program.

Saving the environment around you and preserving the natural luxuries and fores life, that's what eco-tourism is all about. Whether it's about a nature camp or organizing trekking trips towards the unspoit and inaccessible region, one should asway keep in mind not to create any mishap or disturbance in the life cycle of nature. Eco-tourism focuses on local cultures, wilderness adventures, volunteering, personal growth and learning new ways to live on our vulnerable planet. It is typically defined as travel to destination where the flora, tauna, and cultural heritage are the primary attractions. Responsible Eco-tourism includes programs that minimize the adverse effects of traditional on the natural environment, and enhance the cultural integrity of local people. Therefore, in addition to evaluating environmental and culrural factors, initiatives by hospitality providers to promote recycling, energy efficiency, water reuse, and the creation of economic opportunities for local communities are an integral of Eco-tourism.

Importance of Ecotourism in M.P.

In the words of make Twain 'India is a fabulous world of splendor and rags the one country under the sun with an imperishable interest, and the land that all men desires to sec' Ecotourism development has entered an exciting phase in North India. The Indian Ocean, Arabian sea, and Bay of Bengal offer a very large coastline. India is one of the 12 mege bio diverse countries of the world and has rich cultural heritage too. It has vast potential for ecotourism that needs to be tapped for healthy conservation and preservation of nature and bring about economic benefits to the local communities. Ecotourism in India has developed recently for the concept itself is a relatively new one. India has spectacularly attractive natural resources and tourist attraction. India offers enormous diversity in topography, natural resources and climate. There are land-locked mountainous regions, lush valleys and plains, white sandy beaches and islands. Central India has numerous wildlife sancruaries with countless varieties of flora and fauna (Vijay and Chouhan 2007).

REFERENCES

1. UNEP 1995, WTO 1995
2. R.J. Baken and S.Bhagavatula IIM Bangaluru research paper 2010.
3. Vijay kumar and J.S Chouhan ecotourism and practices in India *J Trop Fores 2007 vol.23.*
4. Harish Kumar Khatri: geography of tourism.

Chapter 7

Production of Single Cell Protein from Lignocellulosic Biomass

Anamika Malav and Prahlad Dube

Department of Microbiology,
C.P. University, Kota, Rajasthan

Single cell protein production (SCP) from lignocelluloses biomass presents upcoming technology aimed at providing protein supplement for both human food and animal feeds. Microorganisms such as bacteria, algae fungi and yeast are involved in bioconversion of low-cost carbon feed-stock such as lignocelluloses to produce biomass rich in proteins and amino acids. The cladode of Opuntia ficus-indica (cactus pear), sugarcane (Saccharum officinarum) biomass etc are such lignocellulosic raw material that has a potential for production of SCP in arid and semi-arid regions. This article highlights the current uses of lignocellulosic biomass and the use of cactus pear biomass as potential raw material in SCP production.

Keywords: *Lignocellulose biomass, Raw material, Single cell protein, Yeast.*

Introduction

Single cell protein production usually involves the bioconversion of low-cost carbon feedstocks or low value by-products and wastes into value added microbial biomass. The economics of the process is mainly determined by the cost and availability of the carbon feedstock. Lignocellulosic biomass presents a readily available feedstock or microbial bioconversion which does not compete with feedstocks used for human food. Lignocellulose is the major structural component of woody plants and non-woody plants such as grasses and represents a major source of renewable organic matter. Lignocellulosic biomass is rich in fermentable sugars, high in fibre but low in protein content. The cladodes of *Opuntia ficus-indica*, the prickly pear cactus, is one example of such lignocellulosic biomass. *O. ficus-indica* is well adapted for cultivation in semi-arid regions, with a yield of 10 to 40 tonnes (dry weight) cladode biomass per ha. The cladodes of the spineless *O. ficus-indica* cultivars are suitable as livestock feed during periods of drought. However, because of their low crude protein content (Malanine *et al.*, 2003; Stintzing and Carle, 200), supplementation with other protein sources is desirable for use as a balanced animal feed ration. The cost of supplementation of the cladodes with fishmeal or soybean meal would be prohibitively expensive, however. This review focuses on composition, treatment and bioconversion of lignocellulosic raw material into SCP using different microorganisms.

The first purposeful SCP production originated in Germany during World War I, with the cultivation of baker's yeast, *Saccharomyces cerevisiae*, on molasses and ammonium salts to serve as a protein supplement to replace as much as 60 per cent of the foodstuff Germany had been importing prior to the war (Boze *et al.*, 1992; Giec and Skupin, 1988; Litchfield, 1983). Later, during World War II, Candida utilis was cultivated on diverse waste products from the paper industry to serve as a protein source for both humans and animals (Litchfield, 1979). After World War II, cultivation of fungi in submerged culture for the production of antibiotics led to an investigation of the potential of microfungi as flavor additives to replace mushrooms. In the 1950s some oil industries became interested in the growth of microorganisms on alkanes as a feedstock for SCP production for feed and food (Litchfield, 1980.)

Lignocellulose Biomass

Lignocellulose refers to plant dry matter, so called lignocellulosic biomass. It is the most abundantly available raw material on the earth for the production of biofuels, mainly bio-ethanol. It is composed of carbohydrate polymers (cellulose, hemicelluloses), and an aromatic polymer (lignin). These carbohydrate polymers contain different sugar monomers (six and five carbon sugars) and they are tightly bound to lignin. Lignocellulosic biomass can be broadly classified into virgin biomass, waste biomass and energy crops. Virgin biomass includes all naturally occurring terrestrial plants such as trees, bushes and grass. Waste biomass is produced as a low value byproduct of various industrial sectors such as agriculture (corn stover, sugarcane bagasse, straw *etc.*) and forestry (saw mill and paper mill discards). Energy crops are crops with high yield of lignocellulosic biomass produced to serve as a raw material for production of second generation biofuel; example include grass (*Panicum virgatum*) and elephant grass.

Lignocellulose Biomass as a Raw Material for SCP

Lignocelluloses serve as the major structural component of all plant biomass and represent the major source of renewable organic matter, making it a substrate of enormous biotechnological importance (Malherbe and Cloete, 2002). Lignocelluloses are either derived as a by-product from agricultural products or can be derived from plant biomass grown on non-agricultural or marginal lands. Lignocelluloses are composed of various biopolymers, sugars and chemicals which could be of commercial value. Unfortunately, most lignocelluloses are disposed of as waste. Lignocellulosic feedstocks that have attracted attention for research on SCP production include corn stover, apple pomace, sugarcane bagasse, rice polishing, rice husks, maize cobs, maize fibre and citrus waste. Production of SCP from lignocelluloses is gaining much attention, with the recovery of valuable by-products and simultaneous reduction of the organic load as the chief economic advantages of such processes. Bioconversion of lignocelluloses for SCP production requires various pretreatment methods for the sugars to be hydrolyzed due to the structural and protective role of cellulose and lignin in plants.

Opuntia ficus-indica Biomass as Carbon Feedstock

Crop with the ability to conserve water offer a distinct agricultural advantages in arid and semi-arid regions. One of such crop is the prickly pear cactus (*Opuntia ficus-indica*). The taxonomic genus *Opuntia,* which belongs to the subfamily Opuntioideae, family cactaceae, is a xerophytes consisting of about 200 to 300 species (Stintzing and Carle, 2005). *Opuntia* species contributes in times of drought as life saving crops for both humans and animals. Some species are even naturalized weeds in South Africa and Australia, where the environmental conditions are particularly favourable (Brutsch and Zimmerman, 1993). *Opuntia* is a source of highly digestible energy, water and minerals for animals, but need to be combined with a better protein source to constitute a complete animal feed (de Kock, 2002; Nobel, 2002). For farmers in arid zones, *Opuntia* planting is one solution to alleviate the problem of recurrent droughts. The succulence and nutritive value of *Opuntia* make it a valuable emergency crop, permitting livestock farmers in Brazil, Mexico, South Africa and USA to survive prolonged and severe droughts (de Kock, 2002). Prickly pear cladodes are used as fodder for ruminant animals, but require supplementation with other protein rich forage because of the low protein content of the cladodes. This renders the prickly pear cladodes not suitable for use as sole protein livestock feed. The protein content of the cladodes could be improved by single cell protein production; thus the cladodes have potential of serving as lignocellulosic feedstock for microbial cultivation.

Conclusion

Lignocellulosic biomass is investigated around the globe and many increasing results of this research appeared in literature. Lignocellulosic biomass has the potential of producing good quality and acceptable SCP which may be utilized in various fields. In our study area, sugarcane bagasse, rice polishing, rice husks, maize cobs, maize fibre and citrus waste are available as agriculture byproduct and create environmental problems. This technology gives us an avenue for doing research in this field to solve above environmental problem.

REFERENCES

1. Boze, H., Moulin, G. and Galzy, P. (1992). Production of food and fodder yeasts. *Crit Rev Biotechnol* 12, 65-86.

2. Brustch, M.O. and Zimmermmann, H.G. (1993). The prickly pear (*Opuntia ficus-indica* [Cactaceae]) in South Africa-utilisation of the naturalised weed and of the cultivated plants. *Econ Bot* 47, 154-162.

3. De Kock, G.C. (2002). The use of Opuntia as a fodder source in arid areas of Southern Africa. *In Cactus (Opuntia spp.) as forage*, pp. 101-106.

4. Giec, A. and Skupin, J. (1988). Single cell protein as food and feed. *Nhrung 32*, 219-229.

5. Litchfield, J.H. (1979). Production of single cell protein for use in food and feed. In *Microbial Technology*, 2 edn, pp. 93-155.

6. Litchfield, J.H. (1980). Microbial protein production. *Bioscience 30*, 387-396.

7. Litchfield, J.H. (1983). Single-cell protein. *Science 219*, 740-746.

8. Malherbe, S. and Cloete, T.E. (2002). Lignocellulose biodegradation: Fundamentals and applications. *Rev Environ Sci Biotechnol 1*, 105-114.

9. Malinine, M.E., Dufresne, A., Dupeyre, D., Mahrouz, M., Vuong, R. and Vignon, M.R (2003). Structure and morphology of cladodes and spines of *Opuntia ficus-indica*. Cellulose extraction and characterization. *Carbohyd Polym 51*, 77-83.

10. Nobel, P.S. (2002) Ecophysiology of *Opuntia ficus-indica*. In Cactus (*Opuntia spp.*) as forage, pp. 13-20.

11. Stintzing, F.C. and Carle, R. (2005). Cactus stems (*Opuntia spp*): a review on their chemistry, technology, and uses. *Mol Nutr Food Res 49*, 175-194.

Chapter 8

Conservation of Plants by the Tribes of Bajag, District Dindori, Madhya Pradesh

Ishwar Chandra Parna

Pt. S.N.S. Govt. P.G. College,
Shahdol (Research Centre), Madhya Pradesh

The present paper deal 19 plant species which are conserved by the tribes of Bajag, district Dindori, Madhya Pradesh. Due to destruction of habitat, biotic interference and indiscriminate exploitation of natural plants, many valuable plant species of this area are fast disappearing. Aboriginals conserve these species by faiths, taboos and religious aspects.

Keywords: *Conservation, Tribes, Bajag, Dindori, Sacred forest, Medicine.*

Introduction

Dindori district is situated at the eastern part of Madhya Pradesh. It is lying Between 22° to 23°22" N latitude and 80°35" to 81°58" E Longitude. It touches Anuppur in east, Mandla in west, Umaria in north and Bilaspur district of Chhattisgarh state in south. The district has average rainfall 1400 mm, and temperature 45°C maximum in June and 02°C minimum in December. Several tribals as Baiga, Gond, Kol, Bhariaetc are maintaining their culture and traditions since these culture are influenced by scientific and economic changes, it is therefore essential to study and conserve them before extinct. The tribal utilize large number of plant species in their daily life as food, fodder, fiber, timber, fire wood, broom, gum, medicine *etc*.

The tribes use plant in the forest mainly for preparation of medicines by utilizing natural resources to a small amount it will never harm tribal always very careful to keep the balance between their needs and the conservation of forest. Their knowledge of the use of plants is often kept secret and passed on by verbal

Table 8.1: Plants Conserved by Tribes of Bajag, District Dindori, Madhya Pradesh

Sl.No.	Name of Plant	Common Name	Family	Habit	Plants Part Uses	Reason of Conservation
1.	*Aegle marmelous* (L.) Corr.	Bel	Rutaceae	T	Leaf and Fruit	Sacred plant, the leaves are used to worship lord Shiva.
2.	*Annona squamosa* L.	Sitaphal	Annonaceae	S	Seed and fruit	For fruits and medicine.
3.	*Azadirachta indica* Juss.	Neem	Meliaceae	T	leaf	Plant is an abode of Marhimata,leaves are used in medicine and pest control.
4	*Boswelllia serrata* Roxb. ex Colebr.	Salhen	Burseraceae	T	Wood	The poles of wood are considered auspicious for wedding place.
5	*Buchanania lanzan* Spreng.	Char	Anacardiaceae	T	Fruits and seeds	For fruits and seeds.
6	*Butea monosperma* Lam.	Chhewla (plas)	Papilionaceae	T	Leaves and flower	Leaves are used for thatches, and the flower's are used to worship lord Jagannath.
7	*Calotropis procera* Br.	Akwan (madar)	Asclepiadaceae	S	Flowers and Fruits	The flowers and fruits are used to woship lord Shiva.
8	*Catunaregam spinosa* (Thunb) Tiruv.	Mainhar	Rubiaceae	T	Fruit	Fruits used as vegetable.
9	*Emblica officinalis* Gaertn.	Amla	Euphorbiaceae	T	Whole plant	Sacred plant, worshiped on Amlanavmi.
10	*Ficusbenghalensis*Linn.	Bara	Moraceae	T	Whole plant	Sacred plant.
11	*Ficus religiosa* L.	Pipal	Moraceae	T	Whole plant	The plant is considered an abode of BaramDev (Vishnu).
12	*Madhuca indica* L.F. Gmel.	Mahua	Sapotaceae	T	Whole plant	Sacred plant, flowers used for liquor and the wood is considered anspicious.
13	*Mangifera indica* L.	Aam	Anacardiaceae	T	Whole plant	Sacred plant,The inflorescence is offered to loard Shiva at Mahashivratri.
14	*Ocimum sanctum* L.	Tulsi	Lamiaceae	H	Whole plant	Sacred plant, worshiped by girls for good groom.
15	*Sterculia urens* Roxb.	Kullu	Sterculiaceae	T	Whole plant	The plant is conserved for gum, wood and medicinal use
16	*Terminalia arjuna* Roxb.	Kahua	Combretaceae	T	Whole plant	Sacred plant, The bark is used in medicine.

Sl.No.	Name of Plant	Common Name	Family	Habit	Plants Part Uses	Reason of Conservation
17	*Terminalia bellerica* Gaertn.	Bahera	Combretaceae	T	Fruits	The fruits is used in medicine
18	*Terminalia chebula* Retz.	Harra	Combretaceae	T	Fruits	Roasted fruits is used for the treatment of cough and digestive disorder.
19	*Terminalia tomentosa* Wt. and Arn.	Saja	Combretaceae	T	Whole plant	Sacred plant, dwelling place of boorha Dev.

H: Herb, S: Shrub, T: Tree.

tradition only. A survey of literature indicate that Jain (1963), Khare (2001) have made important contribution in this field.

Materials and Methods

Present Survey was undertaken to collect information from personal interview between author and tribal's of Dindori district. The identification of the collected Specimens was done by using slandered flora.

Results and Discussion

During study period is reported with 19 plant species, which are conserved by the tribals of Bajag, district Dindori for obvious reasons. These tribals organise various occasions and worship plants time to propitiate their gods and goddess. Their beliefs and sentiments are attached to these forests and hence they do not cut or destroy these forests. The botanical name of plants is alphabetically arranged, followed by their local name. All the data obtained as a sequence of present study has been reported (Table 8.1).

Acknowledgement

The auther's thankful to the local tribes and villager for sharing their knowledge with us.

REFERENCES

1. Jain, S.K. (1963) Observation on ethnobotanyof thetribals of Madhya Pradesh.

2. Khare, R.K. (2001) Study of ethnobotany among the tribals of Panna district with special references to biodiversity. Ph.D. Thesis A.P.S. University Rewa (M.P.)

Chapter 9

A Review on the Hepatoprotective Activity of an Important Medicinal Plant *Inula racemosa* Hook. F.

Prachi Sharma

Career Point University,
Kota, Rajasthan

Liver is a vital organ play a major role in metabolism and excretion from the body. Liver plays a major role in detoxification. Any injury or dysfunctioning of it may lead to many complications on one's health. Management of liver diseases is still a challenge to modern medicine. The allopathic treatment has little to offer for the alleviation of hepatic ailments with several side effects. According to Ayurveda, phytochemicals of many medicinal plants posses hepatoprotective activity. Inula racemosa hook f. is one such medicinal plant which shows high hepatoprotective potential. The review discusses the important role of I. racemose in the treatment of hepatic diseases.

Keywords: Liver, Allopathic treatment, Medicinal plant, Inula racemosa Hook. F.

Introduction

Liver is the largest glandular organ of the body which plays a pivotal role to regulate whole metabolic process and homeostasis of the body. It is the most important site of intermediary metabolism and accountable for detoxifying any foreign material and other xenobiotics by converting and excreting waste and toxin. It is considered as one of the most vital organs due to the handling the metabolism of carbohydrates, lipid, protein, secretion of bile, storage of vitamins and production of a variety of coagulation factors. Thus the maintenance of healthy liver is imperative for human health (Pradhan and Girish 2006, Haidry and Malik 2014).

Liver diseases are major global health problem prevalent in developing countries. Liver disease classified as hepatosis (noninflammatory), acute or chronic hepatitis (inflammatory) and cirrhosis or fibrosis (degenerative). It is frequently abused by the environmental toxins, heavy metals, poor eating habits, alcohol, prescription and the counter drug use thus it is damaged and weakened ultimately leads to the hepatitis, jaundice, liver fibrosis and alcoholic liver disease. Liver disease may results in elevated levels of plasma total cholesterol, Low density lipoprotein cholesterol (LDL-C), and Triacylglycerols (TGs) are associated with high risk of atherosclerosis and cardiovascular disease (Dominiczak, 2005).

Although the notable progresses in conventional medical therapy in the last 20 years, drugs available for the treatment of hepatic diseases were limited in efficacy and could have prompted various unwanted side effects when compared to other medical therapies for hepatic diseases which were difficult to handle. In addition, some of these modern hepatoprotective drugs did not protect liver against injury. In response to these factors that limit the use of conventional drugs and efforts were continuously made to identify new sources of agents with hepatoprotective potential (Mamat *et al.*, 2013).

Interestingly, over the last few decades the reputation of use complementary and alternative medicines has increased worldwide due to its therapeutic efficiency and safety, particularly herbal/plant-based therapies, to cure various diseases, efforts are increasingly being carried out by scientists to investigate the hepatoprotective potential of various medicinal plants (Mamat *et al.*, 2013).The plant kingdom is a valuable source of new herbal medicinal agents. In recent years numerous traditional medicinal plants were tested for their hepatoprotective potential in the experimental animals. Traditional medicines and prescriptions with beneficial effects against various pathological conditions have recently paying attention as alternative therapies (Biswas *et al.*, 2014). Several studies have proved the beneficial outcomes of herbal medicine for human health. A variety of molecules have been isolated and their physicochemical and pharmacological properties have been studied. However, compounds and extracts need to be appropriately formulated to facilitate their physiological target and pharmacological activity. Factors such as low permeability and solubility could affect the absorption and delivery of bioactive molecules (Fang and Bhandari, 2011). On the other hand the shelf life of herbal medicine should be evaluated in order to assurance the stability during the period of use. Degradation reactions are enhanced by temperature, humidity, pH, oxygen and light. Herbal medicines are complex mixtures of different classes of chemical compounds, such as carbohydrates, lipids, proteins, and secondary metabolites (Bott *et al.*, 2010).

Inula racemosa Hook. f. (Asteraceae), commonly known as 'pushkarmool', is one such plant which is used as a hepatoprotective in Ayurveda since decades. It grows in the temperate and alpine western Himalayas, and it is common in Kashmir. The roots are widely used locally in indigenous medicine as an expectorant and in veterinary medicine as a tonic. The rhizome is sweet, bitter and acrid in taste with a neutral potency and act as antiseptic, anti-bacterial, anti-fungal, anti inflammatory, analgesic and mild diuretic. It is used in the treatment of contagious fevers, anginapectoris, heart disease and ischemic heart disease. It is also used in cough, hiccup, bronchial asthma, indigestion, flatulence, inanorexia and in fever. Externally, the paste of its roots is used effectively, in dressing the wounds and ulcers as the herb possesses antiseptic property. Also used to boost the appetite2. It is stated in traditional siddha literature under the author Bhava Mishra, 'Bhava Prakash Nigandu'3. Roots of this plant (nagapala) used in liver diseases, rejuvenation and anti ageing. But it has not been explored properly and remains a silent drug in herbal medicine (Gnanasekaran *et al.*,2012).

In present study an effort has been made to review the most prospective herbal plant *Inula racemosa* Hook. F. having pharmacologically most reputable hepatoprotective potential.

Liver is a vital organ play a major role in metabolism and excretion of xenobiotics from the body. Liver injury or liver dysfunction is a major health problem that challenges not only health care professionals but also the pharmaceutical industry and drug regulatory agencies. Liver cell injury caused by various toxic chemicals (certain anti-biotic, chemotherapeutic agents, carbon tetrachloride (CCL4), Thioacetamide (TAA) *etc.*), excessive alcohol consumption and microbes is well-studied. Herbal medicines have been used in the treatment of liver diseases for a long time. A number of herbal preparations are available in the market (Reddy *et al.*, 2014).

Gnanasekaran *et al.*, 2012 evaluated the effect of *in-vitro* hepatoprotective activity of ethanol roots extract of *I. racemosa* on the Chang cell line (normal human liver cells) against carbon tetrachloride induced hepatotoxicity. The cells which are exposed only with toxicant CCl4 showed 42 per cent viability while the cells which were pretreated with extract at concentration of 600 µgmL^{-1} and 300 µgmL^{-1} showed an increase in percentage viability (78 per cent) and the results were highly significant when compared to CCl4 intoxicated cells (Gnanasekaran *et al.*,2012).

Prathyusha *et al.*, 2013, examined the hepatoprotective and curative effect of hydroalcoholic extract of the roots of *I. racemosa* against hepatic ischemic/reperfusion injury in rats. The plant extract at the dose of 200 and 400 mgkg^{-1} produced significant hepatoprotection by decreasing the elevated levels of aspartate transaminase, alanine transaminase, alkaline phosphatase and lactate dehydrogenase. It had been also seen that *I. racemosa* increased the free radicals scavenging activity in the early period of hepatic ischemia/reperfusion injury in rats (Prathyusha *et al.*, 2013).

Another study of Gnanasekaran *et al.*, 2012 was also aimed to evaluate the hepatoprotective activity of the roots of *Inula racemosa* on the Chang cell line (normal human liver cells). The ethanolic extract was tested for its inhibitory effect on chang cell Line. The percentage viability of the cell line was carried out. The cytotoxicity of *Inula racemosa* on normal human liver cell was evaluated by the SRB asasy [Sulphorhodamine B asssay] and MTT assay [(3-(4,5 dimethylthiazole –2 yl)-2,5 diphenyl tetrazolium bromide) assay]. The principle involved is the cleavage of tetrazolium salt MTT into a blue coloured derivative by living cells which contains mitochondrial enzyme succinate dehydrogenase However, the information available on the pharmacological activity of the plant is very limited. Hence, it was proposed to carry out a preliminary in vitro analysis of the hepato protective activity of the plant, which gave promising results (Gnanasekaran *et al.*, 2012).

According to a investigation of Kalachaveedu *et al.*, 2015, Isoalantolactone (IALT), one of the major lactones isolated from the roots of *Inula racemosa* posses high hepatoprotective activity. IALT has been isolated following cold precipitation of its hexane extract and characterized by IR, 1H NMR and 13C NMR spectroscopy. In CCl4 induced liver injury in male wistar rats, at a dose of 100 mg/kg body weight, IALT has brought about significant reduction of SGOT, SGPT and bilirubin levels relative to positive control. The reduction was comparable to that of silymarin (10 mg/kg bw). Histopathological changes corroborated with the altered biochemical marker levels. Reversal of the CCl4 induced pro oxidant status of liver tissue by quenching the generated free radicals could be the possible action mechanism of IALT. This is an interesting observation in view of the HDL-C elevating effect of the parent hexane extract in our earlier study on guinea pigs. This study has demonstrated the hepatoprotective activity of the isolated IALT, a sequiterpene lactone reported with a rich spectrum of biological activity, thus also identifying it as an active principle of the hexane extract of roots of *Inula racemosa* (Kalachaveedu *et al.*, 2015).

Conclusion

Liver damage is the most common cause of mortality and morbidity around the world. It is an epidemic and metabolic disorder. Liver injury treatments are important issue of today's research domain, because of many allopathic drugs and their toxic influence lead to liver damage. Management of liver diseases is still a challenge to modern medicine. The allopathic medicine has little to offer for the alleviation of hepatic ailments whereas the most important representatives are of phytoconstituents. Therefore attention is drawn to the potentials of medicinal plants that have the hepatoprotective ability to reduce or cure liver disorder. The use of herbal plants or their primary and secondary metabolites for curing diseases has long being in continuation since ancient times due to its therapeutic efficacy and safety. Various herbal plants have been investigated for their hepatoprotective potential to treat different types of liver disorder. Numerous herbal plants and formulations are effective in treatment of liver disorder. *Inula racemosa* Hook. F. is one such plant which shows high hepatoprotective potential.This systemic review mainly focused on herbal plants *Inula racemosa* Hook. F. as hepatoprotective in various traditional medicines and explores the herbal plant, isolated active constituent and formulation with hepatoprotective activity.

REFERENCES

1. Pradhan SC, Girish C. Hepatoprotective herbal drug, silymarin from experimental pharmacology to clinical medicine. Indian J Med Res 2006; 124:491-504.

2. Haidry M, Malik A T. Hepatoprotective and Antioxidative Effects of *Terminalia Arjuna* against Cadmium Provoked Toxicity in Albino Rats (*Ratus Norvigicus*). Biochem Pharmacol 2014; 3:1.

3. Dominiczak M H. Lipids and Lipo propteins. In: Baynes JW, Dominiczak MH (Eds) Medical Biochemistry, 2005, (pp 234- 242.). Elsevier mosby. Philadephia.

4. Mamat SS, Kamarolzaman MFF, Yahya F, Mahmood ND, Shahril MS. Methanol extract of *Melastoma malabathricum* leaves exerted antioxidant and liver protective activity in rats. BMC Complementary and Alternative Medicine 2013; 13:326.

5. Biswas A, D'Souza UJA, Bhat S, Damodar D. The hepatoprotective effect of *Hibiscus rosa sinensis* flower extract on diet- induced hypercholesterolemia in male albino wister rats. Int J Med Pharm Sci 2014; 4.

6. Fang Z, Bhandari B. Effect of spray drying and storage on the stability of bayberry polyphenols. Food Chem 2011; 129:1139–1147

7. Bott RF, Labuza TP, Oliveira WP. Stability testing of spray- and spouted bed–dried extracts of *Passiflora alata*. Dry Techno 2010; 28:1255–1265.

8. Gnanasekaran, D., Umamaheswara, C.R., Jaiprakash,B., Narayanan, N., Ravi kiran, Y., Hannah E.(2012). *In-vitro* hepatoprotective activity of *Inula racemosa* roots against CCl4 induced toxicity. *Int. J. Res. Rev. Pharm. App. Sci.* 2(3):578-587.

9. Prathyusha, M., Indala, R., Jagaralmudi, A., Ramesh, K.K. (2013). Hepatoprotective Effect of *Inula racemosa* on hepatic ischemia/reperfusion induced injury in Rats. *J. Bioanal. Biomed.* 5(2): 22-27.

10. Reddy, J.S., Rao, D. and Mallikajuna K. (2014). A review on hepatoprotective activity of some medicinal plants. *International Journal of Innovative Pharmaceutical Research.* 5(2),395-404.

11. Kalachaveedu, M., Kuruvilla, S and Kedike, B. (2015). Hepatoprotective activity of isoalantolactone isolated from the roots of *Inula racemosa* (Hook. F.). *Indian J. Nat. Prod.,*29(1) :56-64.

Chapter 10

Presence of Pb and Cu in Water and Sediment of Shahpur Lake, Bhopal, Madhya Pradesh

Pranita Verma[1], Rashmi Vyas[2] and Rajendra Chauhan[3]

[1]*Department of Zoology, M.L.B College Bhopal, Madhya Pradesh*
[2]*Department of Zoology, P.G. College, Astha, Madhya Pradesh*
[3]*Department of Zoology, M.V.M College, Bhopal, Madhya Pradesh*

Shahpura Lake is a manmade earthen dam where sewage and city garbage disposed. Dam catchment area is completely converted in residential city area. Water is being used for fishing and some agriculture purpose. This is finally meet with Kaliyasot River. Untreated sewage, idol immersion, excessive growth of weeds and siltation is responsible for eutrophication in this lake. In residential city detergent, pesticides, paints and other items are regularly used, in which huge amount of heavy metal discharged in sewer lines and meet to lake. This a study of Pb and Cu heavy metal presence in different sewer line connected with this lake. Water and sediment were collected from five different site to determine Pb and Cu. 9.141, 3.265, 6.128, 0.309 and 11.891 ppm Pb were denoted in water samples whereas 5.321, 8.157, 2.221, 2.451 and 6.361 ppm Pb were found in sediment of that sites. Presences of Cu in these samples are 3.043, 1.395, 0.827, 1.209 and 1.176 in water and 1.035, 0.987, 2.357, 1.215 and 1.364. Heavy metal concentration found in these stations of Shahpura Lake is above upper limit. A sharp management of sewage is needed to maintain these hazardous material mixtures.

Introduction

Shahpura Lake an earthen dam situated near Manisha market Bhopal was made in year 1974-75. It covers about 98 hectare area. This Lake received water from different drainage in which three major are Kotra drainage from North West side, Charimli drainage from northern side and Shahpura drainage from eastern side. Manisha Market Nallah, Campion School Nallah, Panchsheel Nagar Nallah, Chunabatti Nallah releases domestic sewage and rain water to this Lake. These drainages carry direct source of untreated water of municipal, small industrial, household, hospitals *etc*. Overflow of this Lake is flow and meet the Kaliasot River, a stream of Betwa river. This Lake is used for fishing, agriculture, idol immersion, washing, bathing and other religious activities. Due to untreated waste water different sources this Lake covers high concentration of hazardous chemicals and siltation (Giri and Saxena, 2017). This results increased pesticides, toxic chemicals, increased weeds which is due harm to aquatic ecosystem of this Lake. This Lake has good diversity of fish fauna and macro-micro-invertebrates, but due to pollutants diversity is decreasing day to day. Concentration of heavy metal represents the appearance of hazardous pollutants of water (Anu *et al.*, 2011). So, in this study concentration of Pb and Cu in different drainage who release water to this Lake and a sample of Lake Embankment were studied.

Materials and Methods

Water and Sediment samples of W1 (Water Manisha Market Nallah), W2 (Water Campion School Nallah), W3 (Water Panchsheel Nagar Nallah), W4 (Water Chunabatti Nallah), W5 (Water Embankment Station), S1 (Sediment Manisha Market Nallah), S2 (Sediment Campion School Nallah), S3 (Sediment Panchsheel Nagar Nallah), S4 (Sediment Chunabatti Nallah) and S5 (Sediment Embankment Station) were collected in August month of year 2017 for analysis as per following procedure of APHA, (1985) and others.

Water

Sample container was washed and rinsed thoroughly with 5 per cent nitric acid and de-ionized water prior to taken sample from different sites. Sample container was again rinsed with sample at sampling site thrice and taken 500ml sample from at least 1m below surface water. Collected sample were then acidified with 10 per cent HNO_3 and brought to laboratory in ice bucket. Sample were filtered with 0.45μm micropore sieve and kept at 4°C till analysis.

Digestion of Water Samples

100ml water sample was pore in polypropylene beaker. 2ml conc. HNO_3 and 5ml conc. HCl was added and covered with a watch glass. This was heated on heating mental at 90°C until volume was reduced to about 15ml. Beaker was then removed and placed for cool. Watch glass and beaker wall was washed with de-ionised water. Obtained sample was then filtered with Whatman No. 42 paper and diluted with HNO_3 for further analysis (Rohrbough, 1986).

Sediment

Sediment sample were collected from different sites with the help of grab sampler in a dry and clean polyethylene bag. Collected samples were carried out to laboratory and kept that at room temperature for air-dried. Samples were crushed till powdered and filtered in a 160μm sieve. Collected powder sample was kept at -20°C till analysis.

Digestion of Sediment Samples

Sediment samples were digested according to method of Zhelijazkov and Nielson, 1996 with some modification. 1gm dry sample was taken in a 250ml polypropylene tube and mixed 10ml conc. HNO_3. Samples were then heated at 90°C for 45minutes. Temperature was then increased upto 150°C and digested till clear solution obtained. 5ml conc. HNO_3 was then added and digested thrice till found very clear solution and reduced volume about 1ml. tube was washed with de-ionized water to prevent loss of sample. 5ml 1 per cent HNO_3 was then added after cooling and filtered with Whatman No. 42 paper. Obtained sample was then kept for further analysis.

Determination of Elements

Pb and Cu element determination of processed water and sediment samples were carried out in Perkin Elmer AAnalyst 800 Atomic Absorption Spectrophotometer against set WHO parameter (Agemian and Chau, 1975).

Results and Discussion

Water and sediment were collected from five different site to determine Pb and Cu. 9.141, 3.265, 6.128, 0.309 and 11.891 ppm Pb were reported in water samples whereas 5.321, 8.157, 2.221, 2.451 and 6.361 ppm Pb were found in sediment of that sites. Presences of Cu in these samples are 3.043, 1.395, 0.827, 1.209 and 1.176 in water and 1.035, 0.987, 2.357, 1.215 and 1.364. Heavy metal concentration found in these stations of Shahpura Lake is above upper limit. A sharp management of sewage is needed to maintain these hazardous material mixtures. Dayal and Singh in 1994; Dixit and Tiwari in 2008 earlier reported high concentration of heavy metal in water as well as sediment of Shahpura Lake.

Table 10.1: Cu and Pb Concentration in Water and Sediment of different Stations

Sample ID	Analyte (Cu)	Mean	Analyte (Pb)	Mean
Std1	Cu 324.8	[2.5] ppm	Pb 283.3	[5] ppm
Std2	Cu 324.8	[05] ppm	Pb 283.3	[10] ppm
Std3	Cu 324.8	[10] ppm	Pb 283.3	[15] ppm
W1	Cu 324.8	3.043 ppm	Pb 283.3	9.141 ppm
W2	Cu 324.8	1.395 ppm	Pb 283.3	3.265 ppm
W3	Cu 324.8	0.827 ppm	Pb 283.3	6.128 ppm
W4	Cu 324.8	1.209 ppm	Pb 283.3	0.309 ppm
W5	Cu 324.8	1.176 ppm	Pb 283.3	11.891 ppm
S1	Cu 324.8	1.035 ppm	Pb 283.3	5.321 ppm
S2	Cu 324.8	0.987 ppm	Pb 283.3	8.157 ppm
S3	Cu 324.8	2.357 ppm	Pb 283.3	2.221 ppm
S4	Cu 324.8	1.215 ppm	Pb 283.3	2.451 ppm
S5	Cu 324.8	1.364 ppm	Pb 283.3	6.361 ppm

REFERENCES

1. Agemian, H. and Chau, A. S. Y. (1975). An atomic absorption method for determination of 20 elements in the lake sediments after acid digestion. Anal. Chem. Acta. 80: 61-66

2. Anu, Upadhyaya, S.K., Bajpai, A. (2011). Heavy Metal Analysis of Various Water Bodies Located in and around Bhopal, M.P.(India) International Journal of Environmental Science and Development. 2(1): 27-29.

3. APHA, (1985). Standard Methods for the examination of water and waste water. 16th. Ed. APHA, AWWAWPCF, Washington, D.C.

4. Dayal, G. and Singh, R. P. (1994). Heavy metal content of municipal solid waste in Agra, India. Pollut. Res. 13(1): 83-87.

5. Dixit, S. and Tiwari, S. (2008). Impact Assessment of Heavy Metal Pollution of Shahpura Lake, Bhopal, India," International Journal of Environmental Research. 2(1): 37-42.

6. Giri, A. and Saxena, S. (2017). Study of fish diversity of Shahpura Lake, Bhopal, India. World Journal of Pharmacy and Pharmaceutical Sciences. 6(7): 1064-1072.

7. Rohrbough, W.G. (1986). Reagent Chemicals, American Chemical Society Specifications, 7th ed.; American Chemical Society: Washington, DC.

Chapter 11

Microbial Enzymes of Industrial Importance: A Review

Meena[1], Seema Malav[1], Anamika[1] and Prahlad Dube[2]

[1]*Career Point University, Kota, Rajasthan*
[2]*Department of Life Science, University of Kota, Rajasthan*

Microbial enzymes are highly specific and biocatalysis. Today enzymes have become an integral part of the industrial product processing. Microorganisms are favored sources for industrial enzymes due to easy availability, fast growth rate and microbial cells for elevated enzymes production and scientific development in different department. The microbial enzymes have gained recognition global market for their wide spread use in various sectors of industries. In the present paper large number of published and available literature is reviewed.

Keywords: Microbial, Industry, Application, Enzymes, Commercial.

Introduction

Enzymes are biocatalysts produced by living cell to bring about specific biochemical reactions generally forming parts of the metabolic processes of the cells. Enzymes are highly specific in their action on substrates and often many different enzymes are required to bring about by concerted action, the sequence of metabolic reactions performed by the living cell. All enzymes which a have been purified are protein in nature, and may or not possess a non protein prosenthetic groups. The practical application and industrial use of enzymes to accomplice certain reactions apart from the cell date back many centuries and were practiced long before the nature or function of enzymes was understood. Enzymes responsible for bringing about such biochemical reactions become known. It was found that certain microorganisms produce enzymes similar in action to the amylases of malt and pancreas, or to the proteases of the pancreas and papaya fruits. This led to the development of processes for producing such microbial enzymes on a commercial scale. (Dr. Jokichi Takamine. 1894, 1914) was the first person to realize the technical possibility of cultivated enzymes and to introduce them to industry. He was mainly concerned with fungal enzymes, whereas (Boidin and Effront, 1917) in France pioneered in the production of bacteria enzymes about 20 years later. Technological progress in this field during the last decades has been so great that, for many uses, microbial cultivated enzymes. Applications of microbial amylases where both fungal and bacterial enzymes are utilized are in processing cereal products for food dextrin and sugar mixtures and for breakfast foods, for preparations of chocolate and licorice syrups to keep them from congealing, and for recovering sugars from scrap candy of high starch content. Fungal amylases are also used for starch removed for flavoring extracts and for fruit extracts and juices, and in preparing clear starch free pectin. Microbial amylases are used for modifying starch in vegetable purees, and in treating vegetables for canning (Bode, 1954). The microbial enzymes have gained recognition globally for their wide spread uses in various sectors of industries *e.g.* food, agriculture, chemicals, medicine, and energy. Enzyme mediated processes are rapidly gaining interest because of reduced processes are time intake of low energy input, cost effective, non toxic and ecofriendly characteristics. (Li, *et al.*, 2012; Choi, *et al.*, 2015). Moreover, with, the advent of recombinant DNA technology and protein engineering a microbe can be manipulated and cultured in large quantities to meet increased demand (Liu, *et al.*, 2013). Associated driving factors that motivate the use of microbial enzymes in industrial applications are increasing demand of consumer goods needs of cost reduction, natural resources depletion and environmental safety (Choi, *et al.*, 2015). Global market for industrial enzymes was estimated about $ 4.2 billion in 2014 and expected to develop at a compound annual growth rate (CAGR) of approximately 7 per cent over the period to from 2015 to 2020 to reach nearly $ 6.2 billion (2015b: industrial enzyme market). Microorganisms are favored sources for industrial enzymes due to easy availability and fast growth rate. Genetic changes using recombinant DNA technology can easily be done on microbial cells for elevated enzymes production and scientific development (Illanes, *et al.*, 2012). Production of microbial enzymes is a necessary event in the industrial sectors due to the high and superior performances of enzymes from different microbes, which work well under a wide range of varied physical and chemical conditions. Further microbial enzyme used in the treatment of health disorders associated with deficiency of human enzymes caused by genetic problems (Vellard, 2003; Anbu, *et al.*2015). Enzymes, particularly of microbial origin can be cultured largely by gene manipulation, as per the need for industrial applications. Applications of microbial enzymes in food, pharmaceutical, textile, paper, leather, and other industries are numerous and increasing rapidly over conventional methods due to less harm to the higher quality products (Jordon, 1929; Kamini *et al.*, 1999, Gurung *et al.*, 2013).

Microbial Enzymes in Pharmaceutical and Analytical Industry

Microbial enzymes are removal of dead skin and burns by proteolytic enzymes, and clot busting by fibrinolytic enzymes. Nattokinase (EC 3.4.21.62). A potent fibrinolytic enzyme is a promising agent for thrombosis therapy (Sumi, *et al.*, 1987; Cho, *et al.*, 2010). Microbial lipases and polyphenol oxidases (EC 1.10.3.2) are involved in the synthesis of (2R, 3S)-3-(4-methoxyphenyl) methyl glycidate (intermediate for diltiazerm) and 3, 4-dihydroxylphenyl alanine (DOPA, for treatment of parkinson' disease), respectively (Faber, 1997). Cholestrol oxidase (EC1.1.3.6) has also been reported for useful biotechnological applications in the detection and conversion of cholesterol. Putrescine oxidase (EC 1.4.3.10) is used to detect biogenic amines, such as putrescine, a marker for food spoilage (Le Roestill and Prins 2016). Enzymes are indispensable in nucleic acid manipulation for research and development in field of genetic engineering; restriction endonucleases are used for site specific cleavage of DNA for molecular cloning. (Newman, *et al.,* 1995) and DNA polymerases for the DNA amplification by polymerase chain reaction (PCR).

Food Industries

United National Department of Economic and Social Affairs (UNDESA) estimated that world population in predicted to grow from 6.9 billion and food demand is expected to increase by 70 per cent by 2050(http//www.un.org/ waterforlifedecade/food-security.Shtml). These biomolecules are efficiencently involved in improving food production and components, flavor, aroma, color, texture, appearance, nutritive value and good quality food supply issue can be addressed by the application of enzymes in the food industries (Neidleman, 1984). The profound understanding of the role of enzymes in the food manufacturing and ingredient industry have improved the basic processes to provide better markets with safer and higher quality products.(Li *et al.,* 2012). The application of enzymes in food industry is segments into different sectors, such as baking, dairy, juice, production and brewing. Worldwide, microbial enzymes are efficiencently utilized in bakery. The principal application market in food industry to improve dough stability.

Baking Industry

Baking enzymes are used for providing flour enhancement, dough stability, improving texture, volume and color, prolonging crumb softness, uniform crumb structure and prolonging freshness of bread.(2014, Baking Enzymes Market). Bread making is one of the most common food processing techniques globally. The use of enzymes in bread manufacturing shows their value in quality control and efficiency of production. Amylase enzymes are added to the bread flour for retaining the moisture more efficiently to increase softness, freshness and self like. Additionally, lipase and xylanase (EC3.2.1.8) are used for dough stability and condition while glucose oxidase and lipoxygenase added to improve dough strength ening and whiteness (Kuraishi, *et al.,* 1997). Lipases are also used to improve the flavor content of bakery products by liberating short chain fatty acids through esterification and to prolong the shelf life of the bakery products (Andrcu, *et al.,* 1999; Deuter, *et al.,* 1999; Monfort, *et al.,* 1999; Coller, *et al.,* 2000, Kirk, *et al.,* 2002; Fernandes, 2010; Li, *et al.,* 2012; Adrio and Demain, 2014).

Dairy Industry

Dairy enzymes an important segment of food enzyme. Industries are used for the development and enhancing organoleptic characteristics (aroma, flavor and color) and higher yield of milk products. The use of enzymes (proteases, lipases, esterases, lactase, amino peptidase, lysozyme, lectoperoxidase, transglutaminase, catalase etc) in market. Dairy enzymes are used for the production of cheese, yogurt and other milk products (Pai, 2003; Qureshi *et al.*, 2015). Approximately 33 per cent of global demand of cheese produced using microbial rennet.

Beverage Industry

Microbial enzymes are used to digest to cell wall during extraction of plant material to provide improved yield, color, and aroma clearer products (Karlund, *et al.*, 2014). Application of cellulases, amylases and pectinases during fruit juice processing for maceration, liquefaction and clarification, improve yield and cost effectiveness (Kumar, 2015; Garg *et al.*, 2016). Alcoholic amylases may be utilized in the distilled alcoholic beverages to hydrolyze starch to suger prior to fermentation and to minimize or turbidities due to starch. The application of enzymes to hydrolyze unmalted barley and other starchy adjusts facilitate in cost reduction of beer brewing. In brewing development of chill-hazes in beer may be control by the addition of proteases (Okofor, 2007).

Feed Industry

Feed enzymes mainly used for poultry are phytases, protease, a-galactosidases, glucanases, xylanases, α-amylases and poly galacturonases (Walsh *et al.*, 1993; Chesson, 1993; Bhat, 2000; Adrio and Demain, 2014). Feed enzymes are gaining importance as they increase the digestibility of nutrients and higher feed utilization by animals (Choct, 2006).

Polymer Industry

To meet the increased consumption of polymer and growing concern for human health and environmental safety has led to the utilization of microbial enzymes for synthesis of biodegradable polymer. In vitro enzyme catalyzed synthesis of polymer is an environmental safe processing having several advantages over environmental chemical methods (Vroman and Tighzert, 2009; Kadokawa and Kobayashi, 2010).

Paper and Pulp Industry

Increasing awareness of sustainability issues, uses of microbial enzymes in paper and pulp industry have grown steadily to reduce adverse effect on ecosystem. Enzymes are also used to enhance deinking and bleach in paper and pulp industry and waste treatment by increasing biological oxygen demand (BOD) and chemical oxygen demand (COD) (Srivastava and singh, 2015).

Leather Industry

Enzymes are required for facilitating procedure and enhancing leather quality during different stages in leather processing, such as curing soaking, liming, dehairing, bating, picking, degreasing and taining (Mojsov, 2011). The enzymes used in leather industries are alkaline proteases, neutral proteases and lipases. Alkaline proteases are used to remove non fibrillar proteins during soaking, in batting to make leather soft, supple pliable neutral and alkaline proteases, both are used in dehairing to reduce water wastage (Rao, *et al*.1998).

Textile Industry

Enzymes are used to allow the development of environmentally friendly technologies in fiber processing and strategies to improve the final product quality (Choi, *et al.*, 2015). The main classes of enzymes involved in cotton pre-treatment and finishing processes are hydrolase includes amylase, cellulase, cutinase, protease, pectinases and lipase/esterase, which are involved in the biopolishing and bioscouring of fabric antifelting of wood, cotton softening, denim finishing, desizing wool finishing modification of synthetic fibers *etc.* (Araujo *et al.*, 2008; Chen, *et al.*, 2013).

Enzyme in Cosmetics

The applications of enzymes in cosmetics have been continuously increased. Enzymes are used as free radicals scavengers in sunscreen cream, toothpaste, mouthwashes, hair waving and dyeing (Li, *et al.*, 2012). Proteases are used in skin creams to clean and smooth the skin through peeling off dead or damaged skin (Cho, *et al.*, 2007).

Enzymes in Detergents

Enzymes have contributed significantly to the growth and development of industrial detergents, which is a prime application area for enzymes today. Detergents are used in miscellaneous applications as dishwashing, laundering, domestic, industrial and institutional cleaning (Schafer, *et al.*, 2002). Enzymes including proteases, amylases, pectinases, cellulases and lipases used to increase efficiency on stain cleaning and fabric care (Li, *et al.*, 2012).

Organic Synthesis Industry

Enzymes are preferred in industrial chemical synthesis over conventional methods for their selectivity *i.e.* chiral, positional and functional group specific (Schmid, *et al.*, 2001). Catalytic potential of microorganisms have been employed for hundreds of years in the production of alcohol and cheese for industrial synthetic chemistry (Johannes, *et al.*, 2006). Among the enzymes in organic synthesis, lipases are the most frequently used, particularly, in the formation of a wide range of optically active alcohols, acids, esters and lactones (Jaegera and Reetz 1998; Hasan, *et al.*, 2006).

Waste Treatment

The use of enzyme for waste management is extensive and a number of enzymes are involved in the degradation of toxic. The industrial effluents as well as domestic waste contain many chemical commodities, which are hazardous or toxic to the living being and ecosystem. Microbial enzymes (s), alone or in combinations are used for the treatment of industrial effluents containing, phenols, aromatic amines, nit riles *etc.* by degradation or bioconversion of toxic chemical compounds to innocuous products (Klibanov, *et al.*, 1982; Raj, *et al.*, 2006; Rubilar, *et al.*, 2008; Pandey, *et al.*, 2011). The microbial enzymes are also utilized to recycle the waste for reuse *e.g.* to recover additional oil from oil seeds, to convert starch to sugar, to convert whey to various useful products (Kalia, *et al.*, 2001).

Indian Enzymes Market

Around the globe, market is dominated by the food and beverage products and industry that go directly or in directly for human consumption. The biggest challenge in front of fast growing economies such as India is to provide food and health care to even their large population. India, an agriculture based economy, is predicted to grow at 7.9 per cent by 2018 and an attractive market that is opening her doors for industrial enzyme based manufacturing sector. Indian biotech sector accounts 2 per cent of the global biotech market. (Binod, *et al.*, 2013). Recently, Bharat Biotech, Hyderabad-based Pharma Company has developed world first Zika virus vaccine, which is ready for pre-clinical trials demonstrating the "make in india"efforts.

Fat and Oil Processing

In food processing industry the modification of oil and fat is one of the most important areas (Gupta *et al.*, 2003). Fats and oils are essential constituents of foods. The properties of lipids can be transformed by lipases when the position of fatty acid is changed in glycerides and also by exchanging one or more than one fatty acids with new ones. Thus an inexpensive and less needed lipid can be converted into greater value fat. By the use of highly selective phosphplipases, phospholipids in vegetable oils can be eliminated. This process is latest developed and non-toxic to environment (Clausen, 2001).

Oleo-chemical Industries

Immobilized lipases used in oleo-chemical industries to initiate the different reactions (alcoholysis glycerolysis and hydrolysis) used substrate of mix culture. Thus the high productivity and running process will be nonstop with the help of immobilized enzymes. The immobilized enzymes splitting the fat and economically beneficially because there is no need of large investment for thermal energy equipment. In the oleo-chemical industries lipases applications and save the energy in processes of glycerolysis, alcohalysis (Verma, *et al.*, 2012). Modifications with enzymes are beneficial and at moderate conditions reaction can be done (Metzger and Bornscheuer, 2006).

Conclusion

The account given above shows that a lot of work has been done in the field of enzymes which are useful in many industries. The processes for industrial production of microbial enzymes and current industrial uses of enzymes has been presented and the major use of microbial enzymes. Continuously increasing as enzymes have significant potential for many industries to meet demand of rapidly growing population and cope exhaustion of natural resources. Enzymes of microbial origin have significant potential in waste management and consequently in the development of green environment. The enzymes are effectively utilized in many industries for higher quality productions at accelerated rate of reaction with innocuous pollution and cost effectiveness. Microbial enzymes are being used in dairy product, beverage industry, food industry, detergent making, pharmaceuticals, textile industry, cosmetic industry, fuel industry, fat and oil industry, agrochemical, pollution control and in personal care products, waste management. With the help of biotechnology, genetic engineering, and protein engineering are playing very important role to modify the futures of enzymes to increasing their applications in all the industries.

REFERENCES

1. Adrio, JL: and Demain, AL. (2014). Microbial enzymes tools for biotechnological processes. Biomolecules 4(1). 117-139.

2. Anbu, P. *et al.* (2015). Microbial enzymes and their applications in industries and medicine 2014, Biomed Res Int 2015:1-3.

3. Andreu, P. *et al.* (1999). A thermal property of dough's formulated with enzymes and starters Eur. food Res Technology. 209:286-293.

4. Andrio, JL., Demain, AL. (2014). Microbial enzymes tools for biotechnological processes. Biomolecules. 4(1):117-139.

5. Araujo, R. *et al.* (2008). Application of enzymes for textiles fibers processing. Biocatalysis biotechnology 26:332-349.

6. Bhat, MK. (2000). Celluloses and related enzymes in biotechnology. Biotechnological Adv 18:355-383.

7. Binod, P. *et al.* (2013). Industrial enzymes present scenario and perspectines J. Sci. Ind. Res 72:271-286.

8. Bode, H.E. (1954). Enzyme acts as tenderizer. Food Eng., 26; 94.

9. Boindni, A. and Effront, J. (1917). Bacterial enzyme U.S. pat. 1,227,374 and 1,227,525.

10. Chen, S. *et al.* (2013). Cutinase, characteristics, preparation, and application. Biotechnology adv. 31(8):1754-1767 Doi: 10.1016/J biotech adv. 2013.09.005.

11. Chesson, A. (1993). Feed enzymes. Anim. feed sci. technology 45:65-79 Cho SA, Cho JC, Hen SH, (2007). Cosmetic composition containing enzyme and amino acid, Amorepacific Corporation (11/.990): 431.

12. Cho, SA. *et al.* (2007). Cosmetic composition, containing enzyme and amino acid. A more pacific corporation. (11/990):431.

13. Cho, YH. *et al.* (2010). Production of Nattokinase by batch and fed batch culture of bacillus subtillis New biotechnology. 27(4): 341-346. Doi: 10.1016jnbt.2010.06.003.

14. Choct, M. (2006). Enzymes for the feed industry past present and future. World poultry Sci. J. 62: 5-15.

15. Choi, JM. Han, SS. Kim, HS. (2015). Industrial application of enzyme biocatalysis: Current status and futhur aspect. Biotechnology adv 33:143-1454.

16. Clausen, K. (2001). Enzymatic oil. Degumming By a Novel Microbial Phospholiases. Eur. J. Lipid Sci. Technology103:333-340.

17. Dauter, Z. *et al.* (1999). X-ray structure of Novamyl, the five-domain maltogenic' α-amylase from bacillus stearothermophilus: maltose and acarbose complexes at 1.7A' resolution. Biochemistry 38:8385-8329.

18. Faber, K. (1997). Biotransformation in organic chemistry: a textbook Springer. Berlin.

19. Fernandes, P. (2010). Enzymes in food processing: a condensed overview on strategies for better biocatalysis. Enzyme Res Doi: 10.4061/2010/862537.

20. Grug, G. *et al.* (2016). Microbial pectinases: an ecofriendly tool of nature for industries. Biotech. 6(1): 47-59.

21. Gupta, R. N. *et al.* (2003). Lipase Mediated up gradation of dietary fats and oils. Crit. Rev Food Sci. Nutr. 43: (6):35-44.

22. Gurung, N. *et al.* (2013). A broader view: Microbial enzymes and their relevance in industries medicine and beyond. Biomed Res Int. 1-18. Doi: 10.1155/2013/329121.

23. Hasan, s., Shah, AA. Hameed, A. (2006). Industrial applications of microbial lipases. Enzyme microbial technology 39:235-251.

24. Illanes, A. *et al.* (2012). Recent trends in biocatalysis engineering. Bioresour technology 115:45-57.

25. Jaegera, KL. And Reetz, MT. (1998). Microbial lipases form versatile tools for biotechnology. Trends Biotechnology 16(9):396-403.

26. Johannes, T. *et al.* (2006). Biocatalysis. In: Lee S (ed) Encyclopedia of chemical processing. Taylor and Francis New York. Pp 101-110.

27. Jordon, DL. (1929). Red heat in salted hides. J. Int. Soc. Leather trade chem. 13:538-569.

28. Kadokawa, JI. Kobeyashi, S. (2010). Polymer synthesis by cazymatic catalysis. Current opin chem. Biol. 14:145-153.

29. Kalia, VC. *et al.* (2001). Using enzymes for oil recovery from edible seeds. J Sci. Ind. Res 60:298-310.

30. Kalibanov AM. Tu TM, Scott KP. (1982). Enzymatic removal of hazardous pollutants from industrial aqueous effluents. Enzyme Eng. 6:319-323.

31. Kamini, NR. *et al.* (1999). Microbial enzyme technology as an alternative to conventional chemical in leather industry current sci. 76:101.

32. Karlund, A. *et al.* (2014). The impact of harvesting storage and processing factors on health-promoting phytochemicals in berries and fruits processes. 2(3):596-624.

33. Kirk, O. *et al.* (2002). Industrial enzyme application current opin biotechnology 13:345-351.

34. Kumar, S. (2015). Role of enzymes fruit juice processing and its quality enhancement. Adv. Appl. Sci. Res. 6(6):114-124.

35. Kuraishi, c. *et al.* (1997). Production of restructured meat using microbial transglutaminase without salt or cooking J food sci. 62:488-490.

36. Le Rose Hill M Prins A (2016). Biotechnological potential of oxidative enzymes from Actinomyceteria. Doi: 105772/61321.

37. Li, S. *et al.* (2012). Technology prospecting on enzymes: application, marketing and engineering. Comput. Structure biotechnology 2:1-11.

38. Liu, L., Yang, H., HD Shin, (2013). How to achieve high-level expression of microbial enzymes strategies and perspectives. Bioengineered 4(4): 212-223.

39. Metzger, J.O. and Bornscheuer, U. (2006). Lipids as Renewable Resources: Current state of chemical and Biotechnological conversion and diversification. Appl. Microbiology biotechnology 71:13-22.

40. Mojsov, K. (2011). Applications of enzymes in the textile industry. A review in :2nd International congress: Engineering, Ecology and Materials in the processing industry: Jahorina, Bosnia and Herzegovina; Technoloski Fakultet Zvornik, P 230-239.

41. Monfort, A. *et al.* (1999). Expression of LIP1 and LIP2 genes from Geotricum species in baker's yeast strains and their application to the bread making process. J Agric food chem. 47:803-808.

42. Neidleman, SL. (1984). Applications of biocatalysis to biotechnology. Biotech. Genetic eng. Rev. 1:1-38.

43. Newman, M. *et al.* (1995). Structure of Bam HI endonucleases bound to DNA: partial folding and unfolding on DNA binding. Science 269:656-663.

44. Okofor, N. (2007). Biocatalysis: immobilized enzymes and immobilized cells. Modern Ind. Microbiology, biotechnology, P 389. ISBN 978-1-57807-434-0(HC).

45. Pai, JS. (2003). Application of microorganisms in food biotechnology. Ind. J. biotechnology 2:382-386.

46. Pandey, D. *et al.* (2011). An improved bioprocess for synthesis of actohydroxamic acids using DTT (dithiothreitol) treated resting cells of bacillus sp. APB-6 Bioresour Technology 102(11): 6579-6586.

47. Qureshi, MA. *et al.* (2015).Enzymes used in dairy industries. Int J. Appl. Res 1(10):523-527.

48. Raj, J. *et al.* (2006). Rhodococcous rhodochrous PA.34: a potential biocatalyst for acryl amide synthesis. Process biochem. 41:1359-1363.

49. Rao, MB. *et al.* (1998). Molecular and biotechnological aspects of microbial proteases. Microbial Mol. Biol. Rev. 62(3):597-635.

50. Rubilar, O. *et al.* (2008). Transformation of chlorinated phenolic compounds by white rot fungi crit. Rev Environmental sci. technology 38:227-268.

51. Schmid, A. *et al.* (2001). Industrial biocatalysis today and tomorrow nature 409:258-268.

52. Srivastava, N. Singh, P. (2015). Degradation of toxic pollutants from pulp and paper mill effluent. Discovery 40(183):221-227.

53. Sumi, H. *et al.* (1987). A novel fibrinolytic enzymes (Nattokinase) in the vegetable cheese Natto; a typical and popular soybean food in the Japaness diet. Experiential 43: 1110-1111.

54. Takamine, J. (1894). Process of making diastatic enzymes U.S. pat. 225, 820 and 525, 823.

55. Takamine, J. (1914). Enzyme of *Aspergillus oryzae* and the application of its amyloclastic enzyme to the fermentation industry. Ind. Eng. Chem., 6, 824-828.

56. Vellard, M. (2003). The enzyme as drug: application of enzymes as pharmaceuticals. Current opin. Biotechnology 14:444-450.

57. Verma, N. *et al.* (2012). Microbial lipases: Industrial Applications and properties: A review Int. Res J. Biol. Sci. 1(8):88-92.

58. Vroman, I., Tighzert, L. (2009). Biodegradable polymers. Materials 2:307-344.

59. Walsh, GA. (1993). Enzymes in the animal feed industry. Trends Biotechnology 11(10):424-430. http:www.sebi.gov.in/cms/sebi-data/attachdbcs/1453372309087.pdf

Chapter 12

Wetlands of Madhya Pradesh: An Overview

N.R. Suman, A.H. Ansari and R.P. Ahirwar

Department of Botany,
Government Post Graduate College, Damoh, Madhya Pradesh

The present study is based on report of ISA (Indian Space Application) 2011, The Atlas of Wetland of India. Data pertaining to Madhya Pradesh has been analysed at district and divisional level. Datia district represnts least area under wetland, while Khargone has highest area under wetland. At divisional level Ujjain ranked top by sharing 35.5 per cent of land area as wetland area of its constituent districts, while Chambal division represented only 4.15 per cent of area under wetland area.

Keywords: Wetland, District, Division, State.

Introduction

It is interesting to know that, there are nearly 14 x 102 cubic kilometres of water on the planet in which more than 97.5 per cent of the total water in the hydrosphere is deposited in the oceans that cover 71 per cent of the earth's surface. Wetlands are estimated to occupy nearly 6.4 per cent of the Earth's land surface. Nearly 30 per cent is made up of bogs, 26 per cent fens, 20 per cent swamps and 15 per cent flood plains. The amount of fresh water on earth is very small (covers 2.53 per cent of the earth's water) compared to seawater. Of the Earth's fresh water 69.6 per cent is locked away in the continental ice, 30.1 per cent is in under ground aquifers and 0.26 per cent is composed of rivers and lakes. In particular, lakes are founded to occupy less than 0.007 per cent of world's fresh water (Shubhashni 2015).Wetlands are most fragile and diverse ecosystems. These ecosystems harbors numerous species of plants, animals and varied microbes, mostly restricted to particular wetland owing to its specific hydrology, chemistry and geographical situation. Ramsar convention on wetlands (Article 1.1) defines as "area of margen fen, peatland or water, whether natural or artificial permanent or temporary, with water that is static or flowing,fresh, brackish or salt, including areas of marine water the depth of which at low tide does not exceed six meters." overall 1052 sites in Europe: 289 sites in Asia 25 in India and one is M.P. (Bhoj Wetland in Bhopal) : 359 sites in Africa and 79 sites in Oceanic region have been identified as Ramsar sites as wetlands of International importance (Ramsar Secretarat, 2013).

Madhya Pradesh state is located at middle of India in plains between latitude 21°04′N-26.87°N and longitude 74°02′-82°49′ E, Madhya Pradesh state is exactly located in center of the India map so it known as the "Central Region" of India and also known as the "Heart of India" or " Central India". From area point of Madhya Pradesh is the second largest state of India, with Rajasthan being the first, before making "Chhattisgarh" till year 2000, it was the largest state of India in area. Madhya Pradesh is sixth populated state of India.

Table 12.1: Wetlands of Madhaya Pradesh

Sl.No.	Wet Code	Wetland Category	Number	Total Wetland Area	Per cent of Wetland Area	Open Water	
						Post-monsoon	Pre-monsoon
	1100	Inland wetland-natural					
1	1101	Lake/Pond	40	208	0.03	201	43
2	1102	Ox-bow laketlande/ cut-off meander	12	93	0.01	78	-
3	1103	High altitude	-	-	-	-	-
4	1104	Riverine wetland	1	7	-	4	-
5	1105	Waterlogged	20	157	0.02	131	11
6	1106	River/stream	389	315526	38.57	136337	59256
	1200	Inland wetland man-made	462	315991	38.63	136757	59310
7	1201	Reservoir/Barrage	2005	392455	47.97	379592	176276
8	1202	Tank/Pond	15199	64768	7.92	55618	9703
9	1203	Waterlogged	-	-	-	-	-
10	1204	Salt pan	17.204	457223	55.88	435204	185979
		Sub-Total	17666	773214	94.51	57961	245289
		Wetlands(<2.25ha)	44952	44952	5.49	-	-
		Total	62618	818166	100.00	571961	245289

Area under Turbidity levels		
Low	2827	713
Moderate	532712	213784
High	36422	30792

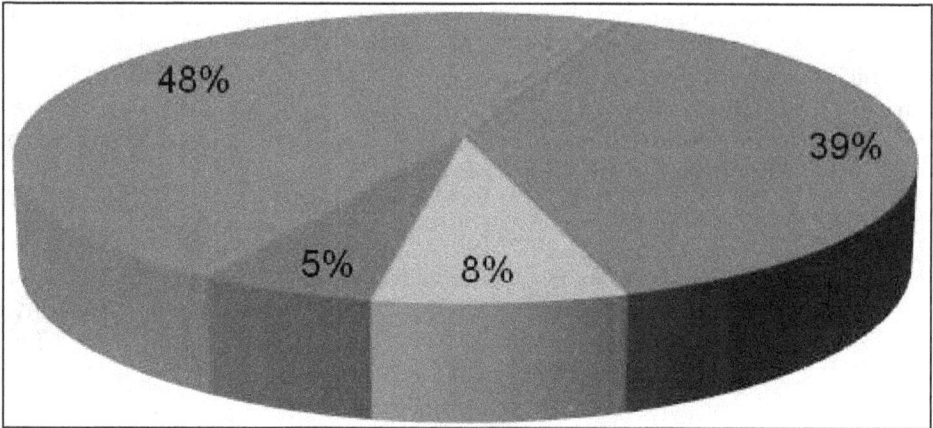

Figure 12.1: Percentage of Various Wetlands in M.P.
Data Source: http://nic.in SAC (2011).

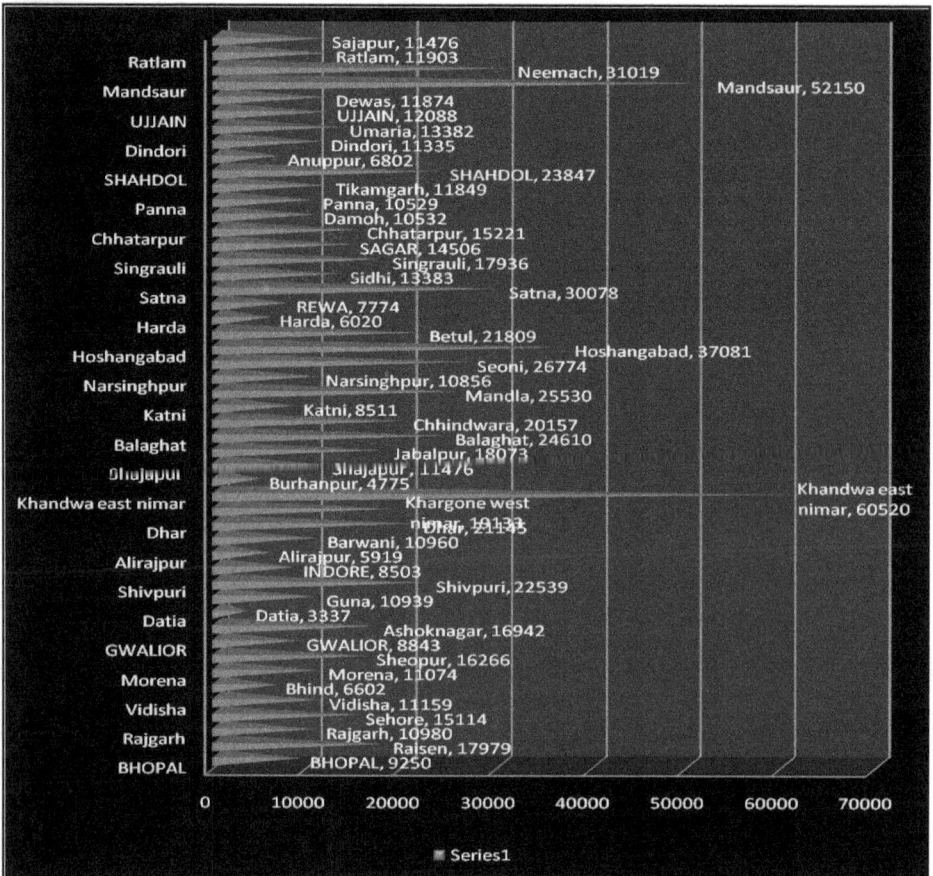

Figure 12.2: Graphic Representation of Wetland Area in different Districts of M.P.
Data Source: http://nic.in SAC (2011).

Extent and Distribution of Wetlands in Madhya Pradesh

In state of Madhya Pradesh total 17666 wetlands have been mapped including small wetlands.Total area under wetland is 818166 hectare.The reservoir/barrage is the major type,accounting (392455 ha), river/streams ranked second with 38.57 per cent share with total area of (31526 ha). The other wetlands types observed are: tank/pond (64768 ha), waterlogged, lake/pond.ox-bow lake and riverine wetlands. Major area of wetland accounting for 48 per cent of total wetland area is represented by small wetlands with less than 2.25 ha.

The diversity, distribution and type of aquatic vegetation is important water quality parameter of water bodies ranging fresh water to saline water.Aquatic vegetation is observed in lake/pond, riverine wetland, tank/pond and reservoir/barrage. The vegetation spread is more during pre-monsoon (62751 ha) as compared to post-monsoon (13379 ha). The open water spread of wetlands is more during post-monsoon (571961) compared to pre-monsoon (245289).Turbidity is moderate in both pre and post-monsoons in a state as a whole.

Inland Wetlands (Natural)

The total open water spread area is 43.27 per cent in post-monsoon, while same is reduced to 18.76 per cent in pre-monsoon, thus there is significant reduction of 24.51 per cent of open water spread area during the sumer season.

Inland Wetlands (Man-Made)

Total open water spread area in man-made wetlands is (435206 ha) in post-monsoon, while same shrinks to (185979 ha). Thus there is 40.67 per cent reduction in open water spread area in man-made wetlands of state.

Top ten wetland rich districts of state contribute 37.02 per cent of total wetlands of Madhya Pradesh. Mandsour ranked the first position by sharing 9.43 per cent of its geographical area and represents 6.37 per cent of total wetland area of state. East Nimar occupied second position by contributing7.97 per cent of geographical area of district and 7.40 per cent of total wetland area of state.Neemuch is third important contributer in terms of wetland area as it contributes 7.27 per cent of district geographical area and 3.79 per cent of total wetland area of state. Fourth important district is Hoshangabad with 5.54 per cent geographical area of district and 4.53 per cent of state's wetland area.Mandla ranked fifth position by sharing 4.40 per cent area of district and 3.12 per cent of total wetland area of state. Satna district is sixth having 4.01 per cent area under wetland and 3.68 per cent of total wetland area of Madhya Pradesh.Seoni,Shahdole,AshokNagar and Singroli ranked seventh,eighth,ninth and tenth position by contributing 3.06 per cent and 3.37 per cent, 4.20 per cent and 2.91 per cent, 3.62 per cent and 2.07 per cent,3.16 per cent and 2.19 per cent geographical area of respective districts and total wetland area of state respectively.

On other hand the following districts represent less then one per cent of geographical area as wetland Datia (0.41 per cent) Burhanpur (0.58 per cent), Alirazpur (0.72 per cent), Bhind 0.81 per cent),Harda (0.74)and Rewa (0.95 per cent). These districts are more water prone areas of state.

Divisional-wise Analysis

There are ten revenue divisions in state of Madhya Pradesh. Divisional analysis of wetland revelead that Ujjain division contributes highest share in terms of state wetland by contributing the land area of 185876 ha which is 22.17 per cent of total werland of state and 35.53 per cent geographical area of its constituent districts. Indore stands second by sharing 142431 ha of land area which is 17.4 per cent of total wetland of state and 22.32 per cent of total geographical area of division. Jabalpur division ranked third by contributing 134511 ha of land area which is 16.44 per cent of state wetland and 19.14 per cent of geographical area of its constituent districts (Figure 12.3). Ujjain division represents highest percentage of land cover under wetland.Rewa is fourth important divison by sharing 69171ha of land which is 8.46 per cent of wetland of state and 11.21 per cent of total geographical area of division.Bhopal division stand fifth by sharing 64482 ha of land area which is 7.88 per cent of state wetland and 11.07 per cent of total geographical area division. This division also have Bhoj international wetland site.

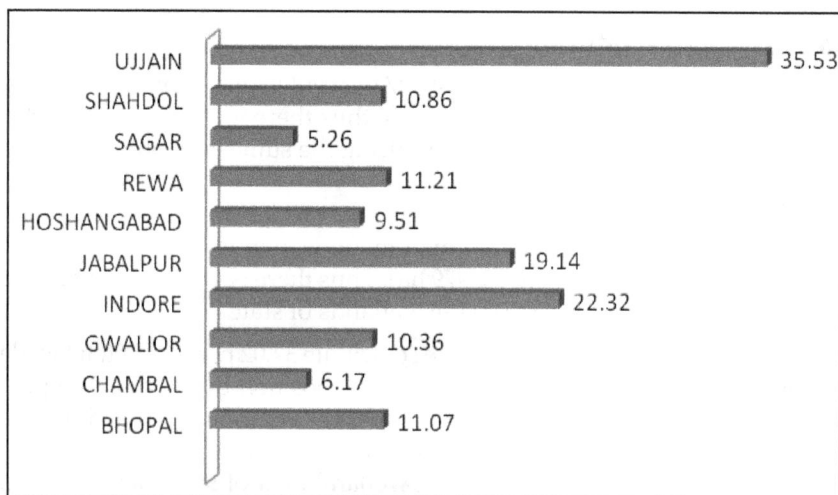

Figure 12.3: Percentage of Geographical Area of Division under Wetland Area (Author's Analysis).

Lowest share of wetland is represented by Chambal division. It contributes just 33942 ha of land area under wetland which is 4.15 per cent of total wetland of state and 6.17 per cent of geographical area of constituent districts. Shahdole is second lowest contributing division which shared 55366 ha land under wetland which is 6.77 per cent of total wetland of state and 10.86 per cent geographical area of division. Sagar is third division having lower wetland distribution. It contributed 62637 ha of land area which is 7.66 per cent of wetland of state and 5.26 per cent of total geographical area of division. Gwalior and Hoshangabad divisions reflected modrate wetland distribution.Both divisions contributed 62600 ha and 64910 ha of land area, 7.65 per cent and 7.94 per cent of wetland of state and 10.36 per cent and 9.51 per cent of total geographical area of divisions respectively.

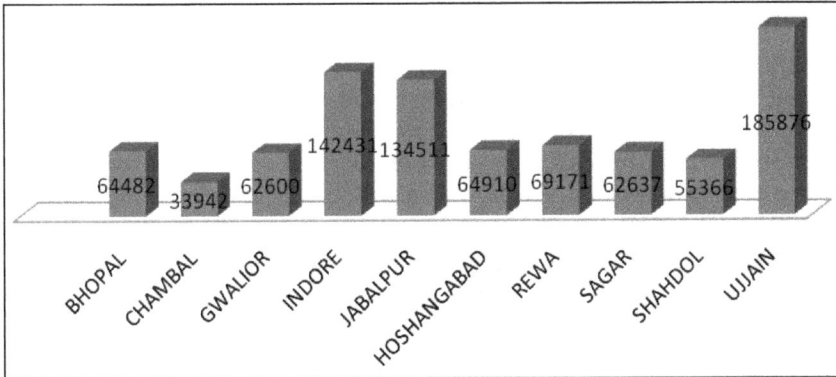

Figure 12.4: Total Wetland Area of Divisions of State (Values in hectare) (Author's Analysis).

Threats to Wetland Areas

The ever increasing population, urbanization, industrialization, developmental forces *etc.* have posed serious threats to very existence of wetlands worldwide. Besides these influxe of sewage uncontrolled aquaculture activities and additions of pollutants have deteriorated the health of wetlands adversely.

Urbanization

Housing projects for accommodating rising population shopping malls, hospitals, and education institutes *etc.* total increase of 12.0 million in last decades.

Population

In terms of percentage of urban population there has been an upward trend with ns increase from 16.6 per cent in 1961 to 27.6 per cent in 2011. The percentage of rural population has decline from 82.0 per cent to 1961 to 68.9 per cent in 2011 in India while in M.P. observed a lower decline in rural population 83.4 per cent to 72.4 per cent from census 1961 to 2011. Madhya pradesh has ranked second with a population of 20069405 in urban areas (www.censusmp.gov.in,2011)

Encroachment

Catchment and adjacent areas of various wetlands are reducing due to illegal encroachment for various purposes such as, agriculture, residential blocks and villeges *etc.*

Waste Disposal

The excessive disposal of domestic, municipal and industrial wastes in water bodies affecting the physico-chemical properties of water along with biotic properties.

Recommendations

Wetlands are cradle of numerous species,good aquafires, groundwater recharger, breeding centres of migratory and native avifauna, controller and regulator of different nutrient cycles and provider of gadzillion ecological services. The very survival of human race is also associated with these wetland. Hence, following measures should be followed for their conservation and protection (T. V. Ramachandra **2001).**

☆ Development of a water quality database, accessible to all users, for analyzing and disseminating information. This can be achieved through:

☆ Exchanging data across departments involved in the program to allow easy accessibility to regularly and continuously onitored data;

☆ Updating technical guidance and water quality maps at regular intervals and indicating quality determinant parameters;

☆ Analyzing and discussing case studies of water quality issues;

☆ Providing spatial, temporal, and non-spatial water quality database systems

☆ Correct non-point source pollution problems and administer the Pollution Prevention Program through environmental awareness programs.

☆ Wetlands require collaborated research involving natural, social, and inter-disciplinary study aimed at understanding the various components, such as monitoring of water quality, socio-economic dependency, biodiversity, and other activities, as an indispensable tool for formulating long term conservation strategies (Kiran and Ramachandra, 1999). This requires multidisciplinary-trained professionals who can spread the understanding of wetland importance at local schools, colleges, and research institutions by initiating educational programs aimed at raising the levels of public awareness and comprehension of aquatic ecosystem restoration, goals, and methods.

☆ Actively participating schools and colleges in the vicinity of the waterbodies may value the opportunity to provide hands-on environmental education which could entail setting up laboratory facilities at the site. Regular monitoring of waterbodies (with permanent laboratory facilities) would provide vital inputs for conservation and management.

☆ An interagency regulatory body comprising personnel from departments involved in urban planning (Bangalore Development Agency, and Bangalore City Corporation, for example) and resource management (Forest department, Fisheries, Horticulture, Agriculture, and so forth), and from regulatory bodies such as Pollution Control Board, local citizen groups, research organizations, and NGO's, would help in evolving effective wetland programs. These programs would cover significant components of the watershed, and need a coordinated effort from all agencies and organizations involved in activities that affect the health of wetland ecosystems directly or indirectly.

References

1. Agarwal, M., 2011. Migratory birds in India: migratory birds dwindling. Nature December.

2. Bassi, N., Kumar, M.D., 2012. Addressing the civic challenges: perspective on institutional change for sustainable urban water management in India. Environ. Urban. Asia 3 (1), 165–183.

3. Bassi, N., Kumar, M.D., Sharma, A., Pardha-Saradhi., 2014. Status of wetlands in India: A review of extent, ecosystem benefits, threats and management strategies.Jour.Hydro.Reg.Studies. (2), 1-19.www.elsevier.com/locate/ejrth.

4. Garg, J.K., Singh, T.S., Murthy, T.V.R., 1998. Wetlands of India. SAC, Indian Space Research Organisation, Ahmedabad.

5. McAllister, D.E., Craig, J.F., Davidson, N., Delany, S., Seddon, M., 2001. Biodiversity Impacts of Large Dams. International Union for Conservation of Nature and United Nations Environmental Programme,Gland and Nairobi.

6. Millennium Ecosystem Assessment (MEA), 2005. Ecosystems and Human Well-being: Wetlands and Water Synthesis. World Resources Institute, Washington, DC.

7. Molur, S., Smith, K.G., Daniel, B.A., Darwall, W.R.T., 2011. The Status and Distribution of Freshwater Biodiversity in the Western Ghats, India. International Union for Conservation of Nature, Cambridge and Gland.

8. Prasad, S.N., Ramachandra, T.V., Ahalya, N., Sengupta, T., Kumar, A., Tiwari, A.K., Vijayan, V.S., Vijayan, L., 2002. Conservation of wetlands of India – a review. Trop. Ecol. 43 (1), 173–186.

9. Ramsar Convention on Wetlands, 2012, September. The Annotated Ramsar List: India. [Brieng Note]. The Secretariat of the Convention on Wetlands, Gland, Switzerland.

10. Ramsar Secretariat, 2013. The List of Wetlands of International Importance. The Secretariat of the Convention on Wetlands, Gland, Switzerland.

11. Space Applications Centre (SAC), 2011. National Wetland Atlas. SAC, Indian Space Research Organisation, Ahmedabad.

12. Subhashini.V.,2015 Restoration,remediation and conservation stratgies of wetlands ecosystems of India.Int.Journ.of Mult. Advanced Resaerch trends

13. T. V. Ramachandra **2001** Restoration And Management Strategies Of Wetlands In Developing Countries ELECTRONIC Green journal http://wgbis.ces.iisc.ernet.in/energy/water/paper/wetland_tvr.htm

14. Verhoeven, J.T.A. Arheimer,B.,Yin,C.,Hefting,M.M.,2006. Regional and global concerns over wetlands and water quality.

15. Verma, M., 2001. Economic Valuation of Bhoj Wetlands for Sustainable Use. [EERC Working Paper Series: WB-9]. Indian Institute of Forest Management, Bhopal.

16. Woistencroft, J.A., Hussain, S.A., Varshney, C.K., 1989.India: introduction. In: Scott, D.A. (Ed.), A Directory of Asian Wetlands.

17. World Wide Fund for Nature (WWF) and Asian Wetland Bureau (AWB),1993. Directory of Indian Wetlands. World Wide Fund for Nature and Asian Wetland Bureau, New Delhi and Kuala Lumpur.Trends. Ecol. Evol. 21 (2), 96–103.\

Annual Report

Environment and Social Welfare Society, Khajuraho India 01 May, 2016 to 30 April, 2017

Important Events in Brief

- ☆ About Environment and Social Welfare Society, Khajuraho India
- ☆ Object of The ESW Society
- ☆ The main branch of Environment and Social Welfare Society
- ☆ International Journal of Global Science Research
- ☆ Appreciation Award to ESW Society
  - ❏ Appreciated by Ministry of Human Resource Development, Govt. of India, New Delhi for Excellence Digital Cashless literacy Campaign.
  - ❏ Appreciated by Maharaja Chhatrasal Bundelkhand Universtiy, Chhaparpur MP for Excellence in Green Festival
- ☆ Collaboration with Institute
- ☆ Appreciation Award by ESW Society
- ☆ Membership/Fellowship of ESW
- ☆ World Environment Day Programme 05, June 2016
- ☆ ISO 9001:2015 Global Certification Services, Bhopal MP
- ☆ Digital Financial Literacy Campaign
- ☆ ESW 4th National Conference 30 and 31 January 2017
- ☆ Scientific Lecture on Envirinment Protection
- ☆ Participation in JAP Mentor Training programme as ESW representative

☆ MOU:

  ❒ MONACHUS, Group of Scientific Research and Ecological Education, Hortensiei Alley, No. 8, 900518, Constanta, Romania

  ❒ Ecological Education/"Dr. FawazAzki" Geological Museum, Syria

☆ Earth Day-2017

☆ Published Book

☆ Science Popularization/Symposia/Seminar/Workshop/Scientific Lecture

☆ Library

☆ National data bases developed

☆ Visual Outputs

About Environment and Social Welfare Society, Khajuraho India

Environment and Social Welfare Society (ESW Society) *Dedicated to Environment, Education and Sciences and Technology entire India since bi-Millennium* is an ISO 9001:2015 certified organization the India. Now it's worldwide known by its impact. It is devoted to Environment, Education, Art and Science and Technology aspect related directly or indirectly to Environment and Social welfare *since Bi-Millennium.* ESW Society has been to develop relationship between Environment and Society envisions the promotion of Education and Sciences among the University, College and School students as well as in the society for Environment and Social welfare as well as Human Welfare. It is registered under the society Act 1973, Government of Madhya Pradesh, India on 31 January 2000. It was affiliated by Nehru Yuva Kendra Sangathan, Ministry of Youth Affairs and Sports, Government of India. It accredited by Madhya Pradesh Jan Abhiyan Parishad, Government of Madhya Pradesh, since 2013. And having NGO-PS, Government of India. NGO Databases.

Object of the ESW Society

1. To establish, arrangement and management all around development in the field of Education and expansions of educational institutions.

2. To develop Ideal morality, Character building in the Children according to Indian tradition and Culture.

3. All around development of the Children. Arrange training programme to establish Self Employment Centre.

4. To organize Seminar for Environmental management, Pollution control, and establish Awareness centre for the same.

5. To make awareness for Social welfare. Check against Animal cruelty and to protect against cruelty and Tyrany.

6. Open animal house for improvement of animal health and provid necessary facility for them.

7. To highlight modern Technology, Computer, Games and Sports, Music, Art, Literature, and various languages Hindi, English, Urdu, and other foreign languages in the field of Education.

8. Establish Research Centre.

The Main Branch of Environment and Social Welfare Society

Godavari Academy of Science and Technology (GAST) Chhatarpur

- ☆ Godavari Academy of Science and Technology, Chhatarpur:
- ☆ Established as prompt Science Academy in India. Goal of Academy is understand the problem, undertaking and solve them by scientific approach of Scientist, Academician or Researcher.
- ☆ Organizing Scientific Session, Science Communication Activity, Seminar, Symposium, Conferences, Workshop, Popular Science Lecture, Interaction with leading scientist, Quiz Contest, Interactive Session on Health and Pollution Control and Career in Biological Sciences are main program.
- ☆ Aim to develop scientific temper among students and social workers. Co-operating with other institute in State and in India as well as Abroad, having similar object and to appoint coordinator of the Academy to act National and International Bodies.

International Journal of Global Science Research

- ☆ International Journal of Global Science Research ISSN E-Version: 2348-8344 (Online) Paris
- ☆ Impact Factor: 1.837 (2015)
- ☆ Frequency: Bi-annual (April and October).
- ☆ Language: English and Hindi.
- ☆ The published papers are made highly visible to the scientific community through a wide **indexing** policy adopted by this online international journal. International Journal of Global Science Research is currently **Indexed in** *ISSN Directory, NISCAIR, Road Directory, Google Scholar, Citefactor, OAJI, DRJI, DAIJ, ESJI, I2OR, IIJIF, CABI UK, ISI, SPARC, AcademicsKeys, Root Indexing, SJIF, JIF, ICI, ISI, IEEE, DOAJ, INDEX COPERNICUS, IEEE, DOAJ, ICI JML, Crossref, COSMOS, Scopus, Thomson Reuters, J-Gate* It is a one-stop, open access source for a large number of high quality and peer reviewed journals in all the fields of Environmental sciences, Bioscience, and Earth Science.
- ☆ Scientists and researchers involved in research can make the most of this growing global forum to publish papers covering their original research or extended versions of already published conference/journal papers/ Review/scholarly journals, academic articles, *etc.*
- ☆ **Abbreviation** Int. J. Glob. Sci. Res.
- ☆ **Journal Website: www.ijgsr.com**
- ☆ **Email: editor@ijgsr.com**
- ☆ If any queries please mail to info@ijgsr.com
- ☆ **PUBLISHED ISSUE:**
  - ☐ **Int. J. Glob. Sci. Res. Volume 3, Issue 2, October 2016**
  - ☐ **Int. J. Glob. Sci. Res. Volume 4, Issue 1, April 2017**

Appreciation Award to ESW Society

☆ Appreciated by Ministry of Human Resource Development, Govt. of India, New Delhi for Excellence Digital Cashless literacy Campaign.

☆ Appreciated by Maharaja Chhatrasal Bundelkhand Universtiy, Chhaparpur MP for Excellence in Green Festival

Collaboration with Institute

☆ The National Academy of Sciences India, Allahabad

☆ Madhya Pradesh Council of Science and Technology, Bhopal

☆ Madhya Pradesh Pollution Control Board, Bhopal

☆ Madhya Pradesh State Biodiversity Board, Bhopal

☆ Maharaja Chhatrasal Bundelkhand University, Chhatarpur MP

☆ Jan Abhiyan Parishad, Chhatarpur, Government of Madhya Pradesh

☆ Government College and School

☆ Cafet Innova Technical Society, Hyderabad, India

Appreciation Award by ESW

National Amazing Godavari Memorial Award (NAGMA)

☆ In the field of "Excellence in Education and Science"

ESW Excellency Award

☆ The purpose of this award is to recognize excellence in leadership involving people, events, programs, projects and teams.

☆ For best co-operation for promotion of Environment and Social Welfare Society at the National and International level.

Recognition Award

☆ ESW Recognition Award For "Valuable Positive Contribution to Environment and Social Welfare Society for Nature conservation

Lifetime Achievements Award

☆ For "performers who, during their lifetimes, have made creative contributions of outstanding environmentalist significance to the field of Environment conservation.

Godavari Academy Impact Award

☆ The purpose of this award is to recognize individuals or teams who have developed, revised and/or implemented a system, tool, process, initiative and/or program within their departments or across the University that had a positive impact.

Godavari Academy Paryavaran Yuva Gourav

☆ For specific contributions to the *Environment and Social Welfare Society* mission and strategic plan

Membership/Fellowship of ESW

To recognize the outstanding academic, environmentalist and scientific contributions of the scientists, the ESW Society awarded the prestigious Membership to the scientists working in different areas of Environment, Education, Art and science and technology, selected from all across the country.

Patron Members

Dr. Kailash Chandra, Scientist 'G', Director, Zoological Survey of India, Ministry of Environment and Forest. Govt. of India, 'M' Block, New Alipore, Kolkata-700053

Life Members

Mr. Arjun Shukla, (FESW) Department of Zoology, Government Model Science College, Jabalpur-482001

Miss. Shivani Rai, RS, Department of Zoology, Government M. H. College of Home Science and Science for Women, Jabalpur-482001

Dr. Mukta Dubey,Guest Lecturer of Political Science, Government College, Rampura -458118

Mr. Shachindra Kumar Dubey, RS, Department of Computer Science and Engineering, Government Engineering College, Gokalpur, Jabalpur- 482001

Dr. Narendra V. Harney, (FESW) Assistant Professor of Zoology, Nilkanthrao Shinde Science and Arts College, Bhadrawati-442902

Dr. Ashwani, (FESW) Lecturer, IGNOU, Bhagini Nivadita College, University of Delhi, Delhi-110007

Prof. Anama Charan Behera, (FESW) Professor of Economics, D. B. College, Turumunga-758046

Mr. Rahul Dev Behera, (FESW) Orissa University of Agriculture and Technology (OUAT), Bhubneswar-751001

Mr. Debashish Sahu, (FESW) Orissa University of Agriculture and Technology (OUAT), Bhubneswar-751001

Mr. Rabindra Nath Padhi, Ex. Deputy Director General, Geological Survey of India, Bhubneswar-751001

Mr. Sandeep Kushwaha, (FESW), Sr., Zoologist, Zoological Survey of India, 'M' Block, New Alipore, Kolkata-700053

Dr. Pragya Khanna, (FESW) Associate Professor of Biotechnology, Government College for Women, Jammu-181101

Dr. **Parvinder Kumar, (FESW)** Sr. Assistant Professor of Zoology, University of Jammu-180006

Dr. **Amit Kumar Bawaria,** Assistant Professor of Chemistry, Government Naveen College, Khadgawan-497449

Dr. **Esha Yadav,** Assistant Professor of Zoology, Janta College, Bakewar-206124

Dr. **Hemlata Pant,** Nematologist, Society of Biological Sciences and Rural Development, Allahabad-211019

Er. **Priyansha Kushwaha,** United College of Engineering and Research, Allahabad-211019

Mr. **Shivam Dubey, RS,** Central Ordnance Depot, Jabalpur-482001

Dr. **Achuta Nand Shukla,** Scientist B, Botanical survey of India, Allahabad-211019

Er. **Saurabh Kushwaha,** Mechanical Engineer, In front of Transformer Raiganj, Gorakhpur-273001

General Members

Mrs. **Anupama Bhargava,** Assistant Teacher of English, P.S. Samadua, Jhansi-284401

Mr. **Vipin Kumar Soni,** Former Guest Lecturer of Chemistry, Govt. Maharaja College, Chhatarpur-471001

Dr. **Sangeeta Chaurasia,** Former Assistant Professor of Zoology, Rajeev Gandhi College, Bhopal-

Dr. **Satyandra Prajapati,** Former Assistant Professor of History, Bapu Degree College, Nowgong 471201

Dr. **Sandeep Kumar Shukla,** Guest Lecturer of Zoology, Govt. College, Seoni-480661

Mr. **Arvind Kumar Dubey,** Guest Lecturer of English, Godavari Academy of Science and Technology, Chhatarpur-471001

Mrs. **Sudha Pauranic,** Former Lecturer, Govt. School, Chhatarpur-471001

Dr. **Vaheedunnisha,** Guest Lecturer of Zoology, Govt. College, Majhauli-486666

Dr. **Deepa Bajpayee,** Technician, Department of Zoology, Govt. Maharaja College, Chhatarpur-471001

Honorable Fellow

Honorable Fellowship Award for "Keen interest in the field of Environmental Sciences" of Environment and Social Welfare (F.E.S.W. Award) to senior most academician and Scientist.

Prof. **Premendu Prakash Mathur,** Vice-Chancellor, KIIT University, Bhubaneswar-751 024

Fellowship of ESW

It also awarded a few Fellowships to the scientists working in different state in collaboration with the scientists in India.

Mr. Arjun Shukla, (FESW) Department of Zoology, Government Model Science College, Jabalpur-482001

Dr. Narendra V. Harney, (FESW) Assistant Professor of Zoology, Nilkanthrao Shinde Science and Arts College, Bhadrawati-442902

Dr. Ashwani, (FESW) Lecturer, IGNOU, Bhagini Nivadita College, University of Delhi, Delhi-110007

Prof. Anama Charan Behera, (FESW) Professor of Economics, D. B. College, Turumunga-758046

Mr. Rahul Dev Behera, (FESW) Orissa University of Agriculture and Technology (OUAT), Bhubneswar-751001

Mr. Debashish Sahu, (FESW) Orissa University of Agriculture and Technology (OUAT), Bhubneswar-751001

Dr. Pragya Khanna, (FESW) Associate Professor of Biotechnology, Government College for Women, Jammu-181101

Dr. Parvinder Kumar, (FESW) Sr. Assistant Professor of Zoology, University of Jammu-180006

ISO 9001:2015 Certification

By Global Certification Services, Bhopal MP **ISO Certification: On 31 July 2016 office of the ESW has been assessed by Global Certification Services, MF-12, A-Block, Mansarovar complex, Bhopal, MP and found to comply the requirements of ISO 9001:2015 Quality Management System for the activities of Environment, Education, Art and All Social work. Certificate issued on 23rd August 2016.**

Digital Financial Literacy Campaign

The Digital India programme is a flagship programme of the Government of India with a vision to transform India into a digitally empowered society and knowledge economy. "Faceless, Paperless, Cashless" is one of professed role of Digital India. As part of promoting cashless transactions and converting India into less-cash society, various modes of digital payments are available. By registered as Volunteer of Godavari Academy of Science and Technology, Chhatarpur wing of Environment and Social Welfare Society, Khajuraho. We help and promote this event in entire India by Social Media. Mr. Margoob Alam Khan, Mr. Deshraj Kushwaha, Mr. Amit Kumar, Mrs. Shivani Shrivastava, Mr. Bharatlal Kushwaha, Mr. Ravi Pal, Mr. Rahul Kumar Ahirwar, Miss. Ritambata Saxena and Mr. Kamta Kachhi were worked for various type of App for cashless transaction.

ESW 4th Annual National Conference 30 and 31 January 2017

Brief Report of ESW IV National Conference on "Impact of Global Warming on Environment, Biodiversity and Ecotourism"

Organized By: Environment and Social Welfare Society Khajuraho-471606 MP, India.

In association with: The National Academy of Sciences India, Allahabad and Maharaja Chhatrasal Bundelkhand University, Chhatarpur MP

Supported By: Madhya Pradesh Council of Science and Technology, Bhopal MP, and **Assisted by:** Godavari Academy of Science and Technology, Chhatarpur, Madhya Pradesh on **30 and 31 January, 2017**

Website: www.godavariacademy.com and www.ijgsr.com

Prof. K.K. Sharma, Former Vice-chancellor Inaugurating
4th ESW National Conference.

A PRELUDE: After the success of 3rd National conference on "Strategy for Human Welfare on Nature conservation and Resource management" The **IV ESW National conference on "Impact of Global warming on Environment, Biodiversity and Ecotourism" 2017** Organized By Environment and Social Welfare Society Khajuraho-471606 Madhya Pradesh, India during 30 and 31 January, 2017 at UNESCO World Heritage Khajuraho, India.

OBJECT: To provide a platform to Vice Chancellors, Educational Administrators, College Principals, Deans, Head of Departments, Professors, Readers, Associate Professors, Assistant Professors, Scientists, Environmentalist, Researchers, and

Young scientists to disseminate knowledge related to Global warming, Environment, Biodiversity and Eco-tourism.

GOAL: Global warming in the major issue in the world. Sustainable Development Goals and its associated targets keeping in mind ESW Society, India President Dr. Ashwani Kumar Dubey has called for action on Quality Education; Clean Water and Sanitation; Climate Action; Life on Land; Peace, Justice and Strong Institutions; Partnerships for the Goals, and Nature conservation to be taken in close coordination with global action on The *Transforming our world: the 2030 Agenda for Sustainable Development.*

THEME: Take some positive steps towards improving our Earth for future generation.

INAUGURAL FUNCTION: The IV ESW National Conference inaugurated on 30 January, 2017 by Chief Guest Prof. K. K. Sharma, Former Vice Chancellor, MDS University Ajmer, Rajasthan. Key note speaker **Dr. A. K. Bhattacharya, MD, National Green Highways Mission, Government of India.** Guest of Honour Dr. Rajesh Saxena, Sr. Scientist, Madhya Pradesh Council of Science and Technology, Bhopal. Doctor S. K. Bhatnagar, Gwalior MP, President Mr. Ravindra Padhi Ex. Deupti Director General, Geological Survey of India, Bhuvneswar, Odissa, Fellow/ Member of Environment and Social Welfare Society Khajuraho, India, Mrs. Vandana Dubey, Managing Director, Godavari Academy of Science and Technology, Chhatarpur, MP and other distinguished guests, participations from various part of India and Two hundred+ listener including media were participated in conference.

Souvenir released with Message of Dr. Kailash Chandra, Director, ZSI, Ministry of Environment and Forest, Govt. of India, Kolkata; Honorable Mahamahim Omprakash Kohli, Governor Govt. of MP; Dr. R. C. Shrivastava, Vice-Chancellor RAU, Pusa, Bihar; Prof. Priyvrat Shukla, Vice Chancellor Maharaja Chhatrasal Bundelkhand University, Chhatarpur; Mr. Chandrakant Patil, Director General, MP Council of Science and Technology, Bhopal MP. Abstract with one hundred fourteen Research Abstract which included from various State of India *viz.* Madhya Pradesh, Uttar Pradesh, Chhattisgarh, Bihar, Maharashtra, Rajasthan, Gujarat, Uttarakhand, Punjab, West Bengal, Chennai. And from abroad Romania.

Prof. K. K. Sharma, Former Vice Chancellor, MDS University Ajmer, Rajasthan addressed on Acousto-informatics: generation, management and application of acoustic database using advanced computing techniques. It may be used for identification, monitoring and management of animal biodiversity.

Dr. A. K. Bhattacharya, MD, National Green Highways Mission, Government of India delivered Key note address on Nature based Ecotourism Scenario.

Dr. Rajesh Saxena, Sr. Scientist, Madhya Pradesh Council of Science and Technology, Bhopal highlighted on Bundelkhand development, livelihood and Ecotourism.

Dr. S. K. Bhatnagar, Gwalior MP discussed on Impact of environment on Cancer.

Released Souvenir by Guest.

Mr. Ravindra Padhi Ex. Deupti Director General, Geological Survey of India, Bhuvneswar, Odissa focused on Climate changes of Earth through ages and its effect on Biosphere.

Dr. Ashwani Kumar Dubey, Executive Director, ESW Society and President and Organizing Secretary of The National conference delivered his presidential address emphasized the role of ESW Society in Global Welfare also focus on annual report of the ESW Society, Khajuraho.

TECHNICAL SESSION: After the inauguration, the scientific session held Fifty + Research paper presented in the technical session.

The general topics covered in the conference will be as under:

Global warming: Temperature increase, Impact of Temperature on Aquatic, Terrestrial and Areal animals, Carbon sinks, Forest and Global warming, Ecology, Ecosystem and its conservation measure, Critical, Natural Disaster, Volcano, Natural calamities, Achieving Global warming Security, Ecosystem services and human welfare. Oxidative Stress and Biomarker, Global warming impact assessment.

Climate change: Climate change, Impact of Food chain and Food web on Human life, Climate change and agriculture, Preventive measure for climate change, Rural Development, Tribal Welfare, Water Conservation, Chemical and Mineral Conservation, Conservation of critical and fragile habitats and corridors, Land degradation and Forest Conservation.

Biodiversity: Diversity, Geo-diversity Biological diversity, Genetic diversity, Species diversity, Ecosystem diversity, Levels of biodiversity, Importance of Biodiversity, Value of biodiversity, Types of biodiversity, Global warming affect Biodiversity, Threats to Biodiversity, Problem on biodiversity, Geo-Biodiversity Conservation and sustainable use of its, Sustainable development, Animal

Behavior and Wildlife Conservation, Endangered, Threatened and Endemic Species Conservation, Strategy for biodiversity conservation, Biodiversity Conservation and Sustainable Management, Conservation issue, endangered species in India, Conservation and promotion of Medicinal plants, Status of Biodiversity, Regional biodiversity.

Eco-Tourism: Tourism, Importance of tourist, Tourist need, Eco-Tourism in India,

Technological Approach Lab to Land: Method and Technique for Global warming, Climate change and Biodiversity conservation, Bio-indicator as a tool of Global warming, Application of bio-technology, Rural bio-technology, Tools and technique: for protection and conservation of biodiversity, Bio-markers with special reference to Global warming, Climate change and Ecosystem management. Role of N.G.O. in Global warming, Climate change and Biodiversity conservation. Pollution, Recycling process of pollutant, Pollution and its monitoring, E-waste and Solid waste management, Eco-Toxicology, Environmental Ethics, Occupational health hazards, Possible solution of Agrochemical and environmental hazards, Environment Conservation and Validation of traditional knowledge.

SCIENTIFIC EXIBITION: An exhibition was arranged along with conference. Researchers got opportunity with delegates and scientist to discuss their needs and publication in the reputed journals.

CULTURAL PROGRAMME: To conserve, promote and develop the Indian's culture, ESW Society arranged cultural event with the national conference.

VALIDICTORY and AWARD CEREMONY ON 31 JANUARY: Prof. Priyvrat Shukla, Vice Chancellor Maharaja Chhatrasal Bundelkhand University, Chhatarpur MP was the Chief Guest, Dr. Satyendra Sharma, Principal, Government PG College, Satna MP, Ex Professor A. L. Dubey, Delhi University, Delhi were Special Guest and Prof. K. K. Sharma, Former Vice Chancellor, Maharishi Dayanand Saraswati University, Ajmer, Rajasthan, was the President of the Valedictory and award ceremony of the Conference. And other eminent scientists were present on this occasion.

Prof. Priyvrat Shukla, Said that the whole world is one family. Dr. Satyendra Sharma, focussed on safe environment and Ex Professor A. L. Dubey, through light on Ecosystem management.

**Released a Book Entitled
"Nature Conservation and Resource Management for Human Welfare."**

AWARD CEREMONY

National Amazing Godavari Memorial Award (NAGMA) in the field of Education and Science awarded to Prof. Priyvrat Shukla, Vice Chancellor Maharaja Chhatrasal Bundelkhand University, Chhatarpur MP

Lifetime Achievement Award: Dr. A. K. Bhattacharya, MD, National Green Highways Mission, Government of India

ESW Excellency Award: Dr. Kailash Chandra, Scientist 'F', Director ZSI, Kolkata WB

Best Paper Oral Presentation Award in each Session awarded to Rakesh Goyal Devi Ahilya University, Indore, Deepak Joshi and Mohd. Danish Govind Ballabh Pant University, Pantnagar, Sandeep Kushwaha, Zoological Survey of India, Kolkata, Arjun Shukla, Jabalpur MP

Best Poster Presentation Award in each session awarded to Shivani Rai, Jabalpur MP and Anuj Kumar Tripathi, Kanpur UP

Young Scientist Award (Below 35 Years) to Sandeep Kushwaha, Zoological Survey of India, Kolkata, West Bengal.

Young Environmentalist Award: Mr. Bibhu Santosh Behera, Bhubaneswar

Environmentalist Award 2016: Dr. Rajesh Saxena, Sr. Scientist, MPCST, Bhopal

Best Scientist Award to Dr. A. K. Pandey National Buro of Fish Genetics Resources, Lucknow

Godavari Academy Impact Award to Prof K. K. Sharma, Former Vice Chancellor, MDS University Ajmer, Rajasthan

Godavari Academy Paryaran Yuva Gourav Award: to Arjun Shukla, Jabalpur MP

ESW Recognition Award: Prof Rashmi Singh Satna and Prof Alka Parashar Bhopal Dr. S. K. Bhatnagar Gwalior Dr. Veena Pandey Nanital, Dr. Prahlad Dube Rajasthan, Prof Satyendra Sharma Satna MP. NGO of South India.

Fellowship of ESW Society Awarded to **Mr. Arjun Shukla**, Madhya Pradesh, Dr. Narendra V. Harney, Maharastra, Dr. Ashwani, Delhi; Dr. Amama Charan Behera, Mr. Rahul Dev Behera, and Mr. Debashish Sahu from Odisha.

Life Member of International Journal of Global Science Research: Mr. Arjun Shukla, Jabalpur; **Miss. Shivani Rai**, Jabalpur and **Mr. Shachindra Kumar Dubey** Jabalpur.

Paryavaran Godavari Award Prof Versha Rani, Uttaranchal, Arvind Prasad Dwivedi, Chitrakoot and Shivani Rai, Jabalpur MP

Certificate of Paper presenter and Participants given by the Chief Guest.

And Mementos presented by Mrs. Vandana Dubey MD Godavari Academy of Science and Technology to all our Guest.

Vote of thanks by Dr. Shivesh Pratap Singh, Secretary BER Chapter, NASI, Chitrakoot

RECOMMENDATIONS:

☆ We summerized recommendations for biodiversity conservation: regional planning; site-scale management; and modification of existing conservation species. We identify major gaps, including the need for more specific, operational examples of adaptation principles that are consistent with unavoidable uncertainty about the future; a practical adaptation planning process to guide selection and integration of recommendations into existing policies; and greater integration of social science into an endeavor that, although dominated by ecology, increasingly recommends extension beyond reserves and into human-occupied landscapes.

☆ Acousto-informatics: generation, management and application of acoustic database using advanced computing techniques. It may be used for identification, monitoring and management of animal biodiversity.

☆ Some planning is required in order to understand and develop the ecotourism Evaluation of existing infrastructure and gap analysis, including existing access opportunities. Assess level of community interest in ecotourism enterprise development;

☆ Develop checklist of requirements for viable ecotourism industry; and

☆ Agreement on community tourism standards (*i.e.* environmental, social, and cultural considerations or concerns)

☆ Nature based Ecotourism Scenario and Sustainable development of Forest management.

☆ **Further research is needed on Impact of pollution on National and World Heritage.**

200+ participants were present out of these General 60 per cent Schedule Caste 10 per cent Schedule Tribes 30 per cent and Women more than 30 per cent overall.

News Gallery

National news paper, Local news paper and electronic channel covered this event promptly.

Foundation Day Celebration 31 January: 17 Foundation day of ESW society celebrated along with National conference at Khajuraho.

Scientific Lecture on Environment Protection: Scientific lecture organized by the ESW Society at Department of Chemistry, Government Maharaja College, Chhatarpur on "Scope of environment safety and sustainable courses in corporate and Industry" on 15, February 2017. Special Lecture given by Dr. Anil Kumar Trivedi, Board Member, School of Earth Science, Central University, Rajasthan. Session chaired by Dr. L. L. Kori, Principal and Dr. P. K. Pateriya. Post Graduation students of Chemistry were present in this event.

MOU: MoU sign between MONACHUS, Group of Scientific Research and Ecological Education, Hortensiei Alley, No. 8, 900518 and Constanta, Romania. Ecological Education/"Dr. FawazAzki" Geological Museum, Syria with Godavari Academy of Science and Technology, Chhatarpur, Madhya Pradesh, India.

Memorandum of Understanding

WHEREAS, the **Group of Scientific Research and Ecological Education/"Dr. FawazAzki" Geological Museum,** hereafter called as **MGSREE/DFAGM** and the **Godavari Academy of Science and Technology** hereafter called as **GAST,** which is headquartered in **Constan a, Romania//Kismin, Syria** and operations in entire world, have come together to cooperate on the following activities:

a) Joint Research and Consultancy

MGSREE/DFAGM faculty and students who are interested in research activities can participate in some of the collaborative projects handled by GAST in liaison with Industry Clients. **MGSREE/DFAGM** may have to identify potential participants with excellent track record in undertaking projects and disclose their expertise so that **GAST** can assign related consultancy works and vice versa.

b) Joint Events

MGSREE/DFAGM and **GAST** can jointly organize the technical events like Symposiums, Conferences, Seminar, Workshops, and Guest Lectures on regular basis to create a platform for discussion forums where the academic and industry experts can exchange their ideas and join hands for collaborative research.

c) Joint Publications

The faculty and students of **MGSREE/DFAGM** can jointly publish research papers with Environmentalist and scientists of **GAST** in Journals published by **GAST** and other leading publishers. **MGSREE/DFAGM** may have to subscribe to all the periodicals published by GAST whereas the faculty and students can avail free publication on annual basis.

WHEREAS, the partner organizations listed above have agreed to enter into a collaborative agreement in which the **MGSREE/DFAGM** will be the lead agency and **GAST** will be the partner agency in this application; and

WHEREAS, the partner organizations herein desire to enter into a Memorandum of Understanding starting the services to be provided by the collaborative for a period of five years from the date of agreement.

Dr. Monica Axini
Executive Manager

Dr. Fawaz Al-Azki
Scientific Manager/Manager
MONACHUS, Group of Scientific Research and
Ecological Education/"Dr. FawazAzki" Geological
Museum Hortensiei Alley, No. 8, 900518, Constanta, Romania

Dr. Ashwani Kumar Dubey
President

Environment and Social
Welfare Society
Vidyadhar Colony,
Khajuraho India

Place: Constanta, ROMANIA/Kismin, SYRIA. Date: 05-April-2017

The Earth Day 22 April, 2017's Campaign
is Environmental and Climate Literacy

The Earth Day Network's mission is to broaden, diversify, and mobilize the environmental movement worldwide to protect the Earth for future generations. The first Earth Day on April 22, 1970, activated 20 million Americans from all walks of life and is widely credited with launching the modern environmental movement. The passage of the landmark Clean Air Act, Clean Water Act, Endangered Species Act and many other groundbreaking environmental laws soon followed. Twenty years later, Earth Day went global, mobilizing 200 million people in 141 countries and lifting environmental issues onto the world stage. More than 1 billion people now participate in Earth Day activities each year, making it the largest civic observance in the world.

Education is the foundation for progress. We need to build a global citizenry fluent in the concepts of climate change and aware of its unprecedented threat to our planet. We need to empower everyone with the knowledge to inspire action in defense of environmental protection. Environmental and climate literacy is the engine not only for creating green voters and advancing environmental and climate laws and policies but also for accelerating green technologies.

Keeping above in our mind we have decided to arrange *Interaction session with Academician, Scientist and Students on the Earth Day* in our Maharaja Chhatrasal Bundelkhand University, Chhatarpur with its collaborates to Environment and Social Welfare Society, Khajuraho.

Event Date: April 22, 2017

Time: 01:00 pm

Place: Conference Hall, Maharaja Chhatrasal Bundelkhand University, Chhatarpur

Published Book "Nature conservation and Resource management for Human Welfare" Editor Ashwani Kumar Dubey Published by Daya Publishing House, A division of Astral International Pvt. Ltd. New Delhi, India. ISBN: 978-93-5124-809-5 (Hardbound) ISBN 9789351248095 (International edition) 2017 http://www.astralint.com/bookdetails.aspx?isbn=9789351248095 pp. 001-110.

About Book: The interest of man in nature goes back to his own origin and history. From stone stage to modern atomic age, there has been a drastic change in the demands of man from the nature ranging from food and shelter to timber, fuel, medicines, pulp and a variety of other products. A sudden explosion in human population and multifarious need of man has put the existence of natural resources to a stake. Majority of original natural resources have been reducing to secondary resources with low productivity hence the extent of wild area is also decreased significantly. Earth is the only place in the universe known to sustain life. Humanity's relationship with the biosphere will continue to deteriorate until a new international economic order is achieved, a new environmental ethic adopted, human populations stabilize, and sustainable modes of development become the rule rather than the exception. Among the prerequisites for sustainable development is the conservation of living resources. the management of human use of the biosphere so that it may yield the greatest sustainable benefit to present generations while maintaining its potential to meet the needs and aspirations of future generations. Thus conservation is positive, embracing preservation, maintenance, sustainable utilization, restoration, and enhancement of the natural environment. Living resource conservation is specifically concerned with plants, animals and microorganisms, and with those non-living elements of the environment on which they depend. Living resources have two important properties, the combination of which distinguishes them from non-living resources: they are renewable if conserved; and they are destructible if not.

Science Popularization/Symposia/Seminar/Workshop/ Scientific Lecture

Researchers/students/teachers/scientists attended the symposia/seminars/ Workshop and Scientific Lectures organized by the ESW Society in different Zone. Several other workshops/sessions were also organized such as Environment and Social awareness programme on World Environment Day, Plant conservation, Health camp, Pollution Awareness programme, Biodiversity conservation, Social awareness programme and Cultural event.

Library

A library service is enriched through subjective books, subscribing more books and by providing facilities of storage, reading room and citation *etc*. Internet facility for educational purpose is also being provided to the students free of cost. The library has been connected to Environment and current event.

National Databases Developed

A huge data base of research papers compiled and published by ESW Society.

Visual Outputs

Annual Reports of last years of ESW Society updated on website.

Plantation at in Mother Teresa Park, Vidyadhar Colony, Khajuraho

Restored Step Wells known as Baolies of Chandra nagar, Bedree, and Rajgarh of Chhatarpur district.

To easy approach for you this ESW Society launch it website on 2013, 31 July at URL http://www.godavariacademy.com

President

Environment and Social Welfare Society,

Khajuraho, India

Index